The Science and Spirituality of Addiction

The Science and Spirituality of Addiction

A Healing Guide for a Broken World

Stuart Morse

FOREWORD BY
Mike Morrell

RESOURCE *Publications* · Eugene, Oregon

THE SCIENCE AND SPIRITUALITY OF ADDICTION
A Healing Guide for a Broken World

Copyright © 2026 Stuart Morse. All rights reserved. Except for brief quotations in critical publications or reviews, no part of this book may be reproduced in any manner without prior written permission from the publisher. Write: Permissions, Wipf and Stock Publishers, 199 W. 8th Ave., Suite 3, Eugene, OR 97401.

Resource Publications
An Imprint of Wipf and Stock Publishers
199 W. 8th Ave., Suite 3
Eugene, OR 97401

www.wipfandstock.com

PAPERBACK ISBN: 979-8-3852-6470-4
HARDCOVER ISBN: 979-8-3852-6471-1
EBOOK ISBN: 979-8-3852-6472-8

VERSION NUMBER 01/06/26

Where noted, Scripture quotations are taken from the NASB® New American Standard Bible®, Copyright © 1960, 1971, 1977, 1995 by The Lockman Foundation. Used by permission. All rights reserved. www.Lockman.org.

Where noted, Scripture quotations taken from The Holy Bible, New International Version®, NIV®. Copyright © 1973, 1978, 1984, 2011 by Biblica, Inc. Used with permission of Zondervan. All rights reserved worldwide. www.zondervan.com.

Where noted, Scripture quotations are from the ESV® Bible The Holy Bible, English Standard Version®, © 2001 by Crossway, a publishing ministry of Good News Publishers. ESV Text Edition: 2025. The ESV text may not be quoted in any publication made available to the public by a Creative Commons license. The ESV may not be translated in whole or in part into any other language. Used by permission. All rights reserved.

For Deb and Aliyah

Contents

List of Illustrations and Tables | ix

Foreword | xi

Acknowledgments | xiv

List of Abbreviations | xv

Introduction | xix

Part One—*Recognition:* **Identifying and Naming Addiction**
1. Inching Towards the Light | 3
2. Fearfully and Wonderfully Made | 17
3. Self Awareness | 28
4. Thinking Outside the Box | 44
5. The Foundations of Addiction Recovery | 62

Part Two—*Understanding:* **Why Addiction Takes Hold**
6. Dopamine | 81
7. Interoception | 98
8. Money | 116
9. Gambling | 132
10. Technology | 148
11. The Devil You Don't Know | 169

12. Opioids and the Placebo Effect | 191
13. Addicted to Love | 205

Part Three—*Integration:* **Developing Sustainable Recovery Practices**
14. Stress | 223
15. An Integral View | 247
16. Growing Up | 267
17. Self Actualization | 283

Bibliography | 305

Illustrations and Tables

Figure 0.1 Vincent van Gogh, *The Starry Night* (1889). Public domain. Image via Wikimedia Commons (MoMA, New York) | xxvii

Figure 3.1 The Operation of SSRI Drugs. Wikimedia Commons | 32

Figure 4.1 The AQAL Matrix | 49

Figure 4.2 Three Twenty First Century World Views | 56

Figure 6.1 Maslow's Hierarchy of Needs. Wikimedia Commons | 84

Figure 6.2 PET Scans of Three Brains. Image source: National Institute on Drug Abuse (NIDA), U.S. Department of Health and Human Services. Adapted from research by Volkow et al. Used under public-domain educational permission | 94

Figure 7.1 The inverted-U. Increased Arousal Correlates with Increased Dopamine and Norepinephrine. Wikimedia Commons. | 99

Figure 7.2 The inverted-U, a Mechanical Analogy. Gyozo Margoczi, used with permission. | 100

Figure 7.3 The Ballmer Peak. xkcd.com, used with permission | 102

Figure 7.4 The Opponent Response to Drugs. Wikimedia Commons | 106

Table 7.1 Feelings Arising When Needs are Met/Unmet | 113

Table 7.2 Some Needs that Must be Satisfied to Promote Optimum Mental Health | 113

Table 10.1 Well known Phenethylamines | 161

Table 10.2 Well known Tryptamines | 162

Figure 14.1 A Personal Sleep Stage Session. EEG Data Recorded Using The Muse Headband. Image © Interaxon Inc., used with permission | 241

Figure 14.2 The Calming Effects of My "Love" Affirmation. EEG data recorded using the Muse headband. Image © Interaxon Inc., used with permission | 245

Figure 16.1 The Rope as a Hierarchy of Holons or Holarchy. Wikimedia Commons | 269

Figure 16.2 Levels of Personal/Cultural Development. © Steve Self. formlessmountain.com 2006 | 273

Figure 16.3 Spiral Dynamics and the Individual Subjective (top-left quadrant.) From *The Integral Vision*, by Ken Wilber, © 2007 by Ken Wilber. Reprinted by arrangement with Shambhala Publications Inc., Boulder, CO. www.shambhala.com | 274

Figure 17.1 The vMeme profiles of Palestine and Israel. Used with permission from admin. of Elza Maalouf's estate Said E. Dawlabani and the Center for Human Emergence Middle East. www.humanemergencemiddleeast.org | 294

Figure 17.2: A Geographical View of vMeme Distribution in Modern Israel. Used with permission from admin. of Elza Maalouf's estate Said E. Dawlabani and the Center for Human Emergence Middle East. www.humanemergencemiddleeast.org | 300

Foreword

WHEN STUART MORSE ASKED me to edit his opus on addiction, spirituality, and neuroscience, I jumped at the chance. I've known Stuart through our overlapping work in personal development and human potential streams, and I've always appreciated his incisive mind and generous spirit. What I didn't expect was how much I would learn from this remarkable book, The Science and Spirituality of Addiction.

As I began reading, I was immediately struck by Stuart's willingness to be vulnerable about his own journey with addiction and recovery. This isn't an academic tome written from an ivory tower of clinical detachment. Instead it's a deeply personal exploration that manages to weave together cutting-edge neuroscience, time-tested wisdom traditions, and raw human experience into something truly unique and valuable.

The biochemistry sections were eye-opening for me. While I've spent years studying consciousness and human development from psychological and spiritual angles, my grasp of the underlying neuroscience has been limited. Stuart's explanations of how dopamine, serotonin, and other neurotransmitters shape our experience and behavior are remarkably clear and accessible. His exploration of conditions like Brunner Syndrome helps us understand how biological predisposition affects behavior, while avoiding both determinism and moral oversimplification.

Perhaps most impressive is Stuart's ability to draw insight from his Christian faith while fearlessly examining difficult questions about human behavior and biochemistry. While my own Christian path tends toward the more progressive and contemplative, and Stuart's theological framework is somewhat more traditional, I found his integration of faith and science to be refreshingly nuanced. He demonstrates how scientific understanding can deepen rather than diminish our appreciation of human spirituality and our need for divine grace.

In exploring the spectrum of addiction, Stuart encourages us to drop our pat condemnations and widen our lens beyond the 'usual suspects' that some of us tend to wax judgmental about, such as drug, sex, and gambling addictions. While not neglecting these, he widens the aperture to include just about all of us in compulsion's lens. His analysis of technology addiction, for instance is particularly timely and insightful. From social media to video games, from smartphones to designer drugs, he helps us understand how modern innovations can hijack our brain's reward systems. His examination of these issues through both contemporary and biblical lenses—particularly his analysis of idolatry then and now—offers fresh insight into age-old human struggles.

The book's exploration of money, power, and status as potential objects of addiction is equally compelling. Stuart helps us see how our economic systems and social structures can enable and even encourage addictive patterns. His integration of Biblical wisdom with contemporary understanding of brain chemistry offers a unique perspective on why Jesus had so much to say about the dangers of wealth and power.

When Stuart ventures into territory more familiar to me—drawing on Integral Theory, Spiral Dynamics, and the work of theorists like Clare Graves, Abraham Maslow, and Carl Jung—he still managed to surprise me with fresh insights. His synthesis of developmental psychology with neuroscience offers new perspectives on why we get stuck in destructive patterns and how we might find our way to greater wholeness.

Particularly fascinating is Stuart's engagement with Steve Peters's "Chimp Paradox" and John Piper's concept of "Christian Hedonism." He shows how understanding our brain's reward systems can actually enhance rather than diminish our spiritual life, helping us grasp why genuine service and worship can be deeply pleasurable without becoming self-serving.

What makes this book particularly valuable is its holistic approach. Stuart recognizes that addiction isn't simply a matter of biochemistry, psychology, or spirituality alone—it's an intricate dance between all these aspects of human experience. By helping us understand these various dimensions and how they interact, he offers hope for those struggling with addiction while maintaining realistic expectations about the challenges of recovery.

I appreciate how Stuart consistently returns to the fundamental human need for connection and meaning. Whether discussing dopamine receptors or Biblical parables, he never loses sight of the fact that at our core, we are beings who long for authentic relationship—with ourselves, with others, and with the divine. His work helps us understand why we so often seek to fill this longing in ways that ultimately leave us feeling more isolated and empty.

The Science and Spirituality of Addiction arrives at a crucial moment in human history, as we grapple with rising addiction rates, increasing social isolation, and profound questions about how technology is reshaping human consciousness and community. Stuart's integrated approach offers valuable insights for anyone seeking to understand these challenges—whether from a personal, professional, or societal lens.

For those struggling with addiction—whether to substances, behaviors, or pursuing wealth and status—this book offers both understanding and hope. For those working in helping professions, it provides a broader context for understanding addictive patterns and potential paths to healing. And for anyone interested in the intersection of science, spirituality, and human development, it offers fascinating insights and thought-provoking questions.

Stuart's work reminds us that while the biochemistry of addiction may be complex, the path to healing ultimately leads us back to elegant truths about human nature and our need for authentic connection. By helping us better understand the science behind our struggles while maintaining sight of our spiritual essence, he offers a valuable contribution to both recovery literature and our broader understanding of human nature.

This is a book that deserves to be read slowly and thoughtfully. Its insights build upon each other, creating a comprehensive framework for understanding both addiction and human development more broadly. I'm grateful to Stuart for writing it, and for allowing me to be part of bringing it into the world.

Michael Morrell

Michael Morrell is an award-winning, best-selling author, collaborator on The Divine Dance: The Trinity and Your Transformation with Fr. Richard Rohr, founder of Wisdom Camp, and a founding organizer of this very justice, arts, and spirituality Wild Goose Festival. He co-parents his two astonishing kiddos in Asheville, NC.

Acknowledgments

I am deeply grateful to my wife, Deb, and my daughter, Aliyah, for their patience, love, and quiet companionship throughout the long process of writing this book. Their presence grounded me in what truly matters.

My thanks to Mike Morrell for his thoughtful developmental editing and his ability to see both the spirit and the structure of the work.

I am also thankful to Kevin Diakiw and Shelley Shadow, whose leadership and guidance in the Addiction Recovery Coach training program at *still-here.ca* helped deepen my understanding of recovery and the human journey.

My thanks also go out to Bob Opperman, the leader of our local Celebrate Recovery ministry, and the men and women of that community for their faith, honesty, and courage. Their stories and fellowship have been a living reminder of hope and transformation.

Finally, I offer this book in gratitude to all those whose journeys through struggle, insight, and recovery continue to illuminate the human spirit.

Abbreviations

5-HT	5-Hydroxy Tryptamine (serotonin)
5HTP	5-Hydroxy-Tryptophan
AA	Alcoholics Anonymous
ACC	Anterior Cingulate Cortex
ADD	Attention Deficit Disorder
ANS	Autonomic Nervous System
AQAL	All Quadrants All Lines
BBB	Blood-Brain Barrier
cAMP	Cyclic Adenosine Monophosphate (a second messenger molecule)
CBC	Canadian Broadcasting Corporation
CBT	Cognitive Behavioral Therapy
cGMP	Cyclic Guanosine Monophosphate (a second messenger molecule)
CNS	Central Nervous System
CT	Computed Tomography
CTE	Chronic Traumatic Encephalopathy
DAT	Dopamine Active Transporter
DM	Dextromethorphan
DMT	Dimethyltryptamine
DSM	Diagnostic and Statistical Manual of Mental Disorders
EDM	Electronic Dance Music
EEG	Electro Encephalogram

EMDR	Eye Movement Desensitization and Reprocessing
EPP	Endplate Potential
ESV	English Standard Version
fMRI	Functional Magnetic Resonance Imaging
FOMO	Fear Of Missing Out
GABA	Gamma Aminobutyric Acid
GPS	Global Positioning System
IDF	Israeli Defense Forces
IIT	International Intensive Training
LCD	Liquid Crystal Display
LED	Light Emitting Diode
MAO	Monoamine Oxidase
MAOI	Monoamine Oxidase Inhibitor
MDMA	Ecstasy (the drug) Methylenedioxy Methamphetamine
MRI	Magnetic Resonance Imaging
MSG	Monosodium Glutamate
NASB	New American Standard Bible
NIV	New International Version
NMDA	N-Methyl-D-Aspartate (a glutamate receptor)
NREM	Non-Rapid Eye Movement sleep
NVC	Non-Violent Communication
PAC	Political Action Committee
PEA	Phenethylamine
PET	Positron Emission Tomography
PFC	Prefrontal Cortex
PIHKAL	Phenethylamines I have Known and Loved (book by Alexander Shulgin)
PNS	Parasympathetic Nervous System
REM	Rapid Eye Movement sleep
SDT	Spiral Dynamic Theory
SSRI	Selective Serotonin Reuptake Inhibitor
$T_1R_{1/2/3}$	Taste Receptors 1, 2, and 3
$TAAR_1$	Trace-Amine Associated Receptor 1

THC	Tetrahydrocannabinol
TIHKAL	Tryptamines I have Known and Loved (book by Alexander Shulgin)
UCSF	University of California San Francisco
VCH	Vancouver Coastal Health
VGCC	Voltage Gated Calcium Channel
WWII	World War 2

Introduction

BRITISH TV SHOWS ARE among the most engaging you can find. They distinguish themselves from the pack through storytelling, character development, excellent acting and their focus on substance rather than glamor. Thanks to the Internet, viewers around the world can binge-watch their favorite productions through streaming services like BritBox and Acorn TV. While you are enjoying the show, you might notice the frequency with which a glass of wine or beer comes into frame. This emphasizes the important role alcohol plays in the social fabric of British culture—the environment in which I was raised. As a teenager I spurned alcohol's influence in my life, seeing it as shallow and unnecessary, but as time passed, I realized it was an almost unavoidable component of British social life. Alcohol contributes positively to the relational endeavors of the healthy individual. While alcohol is merely one of many substances of abuse, it will remain the focus of this introduction.

Going beyond the role of alcohol as social lubricant it also stands as a signifier of manliness. As a male, your ability to consume large quantities of the stuff while remaining functional adds greatly to others' perception of your masculinity, while the term "he can't handle his drink" is among the worst insults. Drinking also starts early in England. At Christmas and on occasion throughout the year my high school friends would smuggle hard liquor onto the school grounds where it was consumed surreptitiously. If you looked old enough it was also common to begin your drinking career in public establishments long before the legal drinking age of eighteen.

In my late teens I had an ambivalent relationship with alcohol (I couldn't decide if I liked it) but in the first year of university I began to notice its calming influence. Alcohol's soothing presence helped fortify me against the academic rigors of a four-year degree in Computer Science. After a long day of classes followed by hours of homework, I experienced a feeling of disquiet that was completely extinguished by a pint or two of my favorite adult libation. This became my nightly practice, a pursuit that supported my psyche for the remainder of the program.

Out in the working world I was introduced to the concept of the "liquid lunch." This was usually a menu option on Fridays or during holiday seasons, but it was also pursued by some on a more regular basis. I still remember my boss consuming three imperial pints (1.7 liters or sixty oz) of beer (without food) over the lunch hour and returning to a functional afternoon of work at the office. This was impressive and relatively normal for the culture of the time. Yes, there were a few weekday hangovers involved in that lifestyle, but they were seen as badges of honor and humorous adjuncts to the workday world.

How Much Is Too Much?

Could these behaviors be the markers of addiction? One way to assess this might be to measure consumption against the health guidelines of your local jurisdiction. Having lived in Canada for over thirty years, it was Canada's Low-Risk Alcohol Drinking Guidelines (LRDGs) issued in 2011 that first caught my attention[1]. They claim that for men, consuming fifteen standard drinks a week is safe, while for women the limit is ten. In this context a standard drink is twelve oz (355 ml) of beer, five oz (148 ml) of wine or 1.5 oz (44 ml) of spirits.

As middle age approached, I was still managing my evening feelings of disquiet by consuming half a 750 ml bottle of wine (2.5 standard drinks) six nights a week, which falls within the 2011 healthy range for men. I asked my doctor for his opinion on this to which he responded, "that's nothing, you should see how much some of my patients drink." Perhaps he was referring to people like a gardening acquaintance of mine. I noticed the smell of alcohol on his breath as we rode the bus together one morning, so I asked him how much he had already consumed that day, "three beers" was his reply (it was 7:45 am.) Being somewhat surprised by his response I asked what his normal daily consumption

---

1 Health Canada, "Low-risk", para. 3.

might be, to which he replied, "a couple of cases, twenty four." That amounts to a total of 168 standard drinks per week.

Could consuming of twenty four cans of beer a day be regarded as an addiction? Probably, but this is likely the wrong metric to use in making this determination. If we define addiction as any *habitual* behavior that brings harm to the self or others, addiction might involve far smaller amounts. This reframes the threshold for addiction from a measure of volume to the question of why the user is drinking in the first place—why is their drinking a habitual behavior and what needs does it meet? Consuming half a bottle of wine a night might not result in the existential crises of familial collapse, loss of employment or jail time, but it does incur a cost.

Reassessing the Risks of Alcohol Consumption

Until recently it was believed that moderate alcohol consumption bestowed certain health benefits on the imbiber. This led some to fear that abstinence could harm their overall health, an idea often used by the psyche to maintain alcohol's hold on their lives. In 2023, however, the World Health Organization determined that *no* level of alcohol consumption is safe[2] and that alcohol always demands payment in more than just financial terms. In the same year, the Canada Centre on Substance Use and Addiction published a new set of alcohol consumption guidelines[3] indicating that consuming more than two standard drinks per week increases a person's risk of developing several types of cancer and more than seven drinks per week also raises the risk of heart disease and stroke.

Alcohol's financial cost alone is steep. Over the past ten years, the cost of a glass of wine at a restaurant in my hometown has more than doubled. It now costs what two bottles did at the liquor store ten years ago. At that price point even moderate consumption can substantially impact an individual's financial bottom line.

Alcohol's impacts on those who use it to function are multi-faceted. On the familial front, by the time alcohol use has brought an end to a spousal relationship it has already inflicted its harmful influence on all parties involved for an extended period. But the destruction of the

---

2 World Health Organization, "No level," para. 1.
3 CCSA, "Canada's Guidance," para. 3.

relationship is by no means a foregone conclusion if those involved are open to exploring solutions.

Relational damage is not the only indicator of familial harm. In 2017 Roland Orzobal (co-founder of the band Tears for Fears) lost his wife Caroline to alcohol-induced dementia at the age of fifty four. Concerning this tragedy he wrote:

> We'd been drinkers, wine drinkers, and in England it's kind of accepted as being okay. I'm drinking a bottle-and-a-half of wine a night, you know, that's just me. Now, Caroline was matching me. And she was quite a small woman. So, her problem was largely due to alcohol[4].

The threat to their relationship was not alcohol-induced strife or "hitting rock-bottom" in terms of living conditions; it was in the consistent, jointly consensual consumption of a toxic substance. Her condition inspired Orzobal to pen the title track of the 2022 album "The Tipping Point," a song that explores a person's journey from being mostly alive to mostly dead. As a man, with a larger frame, Orzobal avoided the worst of alcohol's toxic effects. This highlights the additional risk inherent in drinking for women.

The magnitude of alcohol's personal and societal impact stems, in part, from the belief that a user needs to hit "rock-bottom" before they turn their lives around. Depending on a drinker's self-awareness skills, a complete collapse might be required to convince them that something needs to change, but making rock-bottom the de-facto threshold for triggering outside involvement disqualifies many early interventions. Because of this, friends and family often feel excluded from offering support when it is most needed. For this reason, the addiction recovery community is reclassifying the threshold for intervention from a question of "have you hit rock-bottom yet?" to "does your habit engender harms?"[5]. This book takes the second path with the goal of helping all those who dull their pain using substances and behaviors, not just those that require residential rehabilitation. Whether you consume three drinks per day or twenty three, the same causes and solutions apply.

---

4 CBS, "Tears for Fears," para. 39
5 Ganz, "Unpacking."

My Canadian Life

Years after emigrating to Canada and marrying my wife of thirty seven years, I found myself in a technological backwater on the employment front. This prompted me to pursue a master's degree in computer science, specializing in the field of Artificial Intelligence. This was before vast amounts of knowledge from the World Wide Web had been incorporated into language models like Open AI's ChatGPT, Google's Gemini 2, and Meta's LLaMA. When I was contemplating the particulars of AI in the mid-1990's, it was the study of knowledge representation, knowledge elicitation, and machine learning. The goal was to reproduce the machinery of human rationality using computer software.

Far from being merely technical in nature, this line of study was also a philosophical endeavor. The goal was to understand the structures of conscious reason and build software systems that mimicked their biological analogues. Ultimately this approach was superseded in popularity by the probabilistic approach of today's neural networks. It did, however, spur my interest in the operation of the human brain and how its influence directs our lives. During this period, I was also schooled in research methods, a skill that I was able to use in a cross-disciplinary context while compiling the chapters that follow.

In addition to my academic pursuits, I was also heavily involved in my religious community. I saw no incongruity between my moderate daily consumption of wine and my faith-based activities. As a member of a Christian lay community, I had plenty to do in the way of managing, teaching, preaching, running seminars and leading the occasional international speaking engagement. My alcohol consumption was well below that required to compromise my daily performance. What did prick my conscience, however, was the knowledge that I was using alcohol to manage feelings that stemmed from childhood trauma. I realized that rather than being a solution to my hurts, alcohol was preventing my healing. Instead of accepting my drinking as a normal feature of life (it was after all within the 2011 health guidelines) I decided I would be better off without it. Wine is no longer something I lean on for support, but I still enjoy the occasional glass in social situations. My wife's encouragement was an invaluable influence in making this transition.

The Twelve Step Program

I began my healing journey by assessing the available recovery tools. Alcoholics Anonymous and Narcotics Anonymous with their Twelve Step approach had the highest profiles, but I was struck by the story of nurse Byron Wood, a worker at Vancouver Coastal Health[6] for whom these approaches were inapplicable due to their reliance on a faith he did not possess. Wood was diagnosed with an addiction to alcohol and opioids when he was hospitalized following a psychotic break in 2013. While he kept his job, his employer required him to receive treatment for his addiction and assigned him to the care of a VCH-approved doctor. As part of his treatment his doctor arranged for him to attend a Twelve Step Program run by Alcoholics Anonymous. Wood attended some sessions but quit due to the clash between their theocentric approach and his atheist beliefs. VCH fired him in response, as the only treatment they offered involved the Twelve Steps. In Wood's words:

> Six of AA's Twelve Steps directly refer to God or a higher power, including one that requires members turn their will and lives "over to the care of God." The Twelve Steps are a religious peer support group, not a medical treatment. They shouldn't be imposed on anyone. When you're a medical doctor, and you specialize in only one condition, and the only treatment that you offer for that condition involves God, you shouldn't be practicing medicine.

Wood filed a human rights complaint alleging that his dismissal was a case of discrimination against him due to his religious beliefs. The courts agreed with Wood. While many details of his settlement are confidential, VCH employees are no longer required to attend AA and similar programs if that approach to treatment conflicts with their religious or non-religious beliefs. They may instead request help from secular support groups like SMART Recovery or LifeRing Secular Recovery. A God-centric approach to addiction recovery does not work for everyone, but many atheists and agnostics still benefit from the Twelve Step approach by opening their minds in creative ways.

As a practicing Christian I had no issues with the Twelve Step approach, but I was curious why the standard mandatory treatment for workplace addiction issues in Canada has long been Alcoholics Anonymous. Evidently Government institutions have consistently judged it to

---

6 Lindsay, "Atheist Nurse."

be the most effective solution when compared with the alternatives. But why is the belief in God, or a higher power, an effective recovery tool for many addicts? This book is, in part, my answer that question. As shown in Wood's case, the Twelve Step approach is ineffective without an openness to a higher-power's influence on the part of the addict, and as God speaks most authoritatively through the words of the Bible, for most Christian and Jewish believers, we will look to this book to guide our inquiry.

Where Do We Start?

In assessing potential solutions to our addictive behaviors, it helps to consider why addictions come into being in the first place. This consideration is not one of mathematical precision, but it can be bolstered using cognitive tools like the *metaphor,* concerning which the Spanish philosopher and essayist José Ortega y Gasset wrote:

> The metaphor is probably the most fertile power possessed by man.

One metaphor that stands as a clarion in this discussion is the concept of *psychic turbulence*—turbulence as it relates to the mind. This term is often used in psychological, philosophical or literary contexts to illustrate the emotional and mental disturbances an addict seeks to address by using. When viewed from this perspective, recovery emerges as the process of calming the turbulence. What follows is an account of how psychic turbulence might manifest in a person's life.

At birth we begin our journey on the road of life as smooth vehicles over which the air glides effortlessly, maintaining a stable, uninterrupted flow. If we experience the optimal set of life conditions as we develop, we will travel down this road with little resistance and achieve our highest capabilities at each stage of the journey. If, however, we suffer psychic injury on our journey, our mental bodywork becomes battered or torn. Because of this damage we no longer cut effortlessly through life but encounter resistance to our forward motion. This manifests as drag through the formation of swirling vortices which split from life's flow in response to the injuries incurred. Some people are born with surface irregularities, with similar effect. These include those with conditions like anxiety, dyslexia, ADD, and autism. Further psychic injury exacerbates these conditions and increases the effects of drag.

In addition to reducing our forward velocity, these hurts impact our stability; we must invest more energy to stay on track. If we attempt to move at the same speed as our streamlined peers, we are likely to lose control of our vehicle and end up in the ditch. While the effects of mental turbulence are apparent to us, we are ignorant of their cause or resolution. Even expert rehab counsellors find our situation challenging. Psychic turbulence is an obstacle which (at first) derails the recovery of most, no matter what interventions are used. Multiple interventions may also be required to calm the pandemonium.

The complexity of turbulent flows is illustrated by an anecdote from my university days. One of my computer science professors was asked by an employer to update a software package that simulated the basic attributes of airflow. After he had demonstrated his ability to enhance the software, they said "Good. Now we want you to modify the code to simulate turbulent conditions." After working on the task for six weeks he began spending less time working on the program and more time looking for a new job. He quit shortly afterwards. Rather than an indication of incompetence on the part of my professor this was a testament to the complexity involved in understanding and modelling turbulence.

The eminent Twentieth Century physicist Werner Heisenberg was also familiar with the challenges presented by the physics of atmospheric turbulence. On his deathbed he reportedly quipped:

> "When I meet God, I am going to ask him two questions: Why relativity? And why turbulence? I really believe he will have an answer for the first."

By this he implied that turbulence seems so chaotic and inscrutable that even God might struggle to explain it. There is, in fact, a million-dollar reward available for the first person who solves this mathematical conundrum[7]. The world's most accomplished scientists and rehab counsellors have both been unable to fully model the complexity of turbulence in their respective fields. In each case the observable features are complex, chaotic and unpredictable. And yet Vincent Van Gogh, through intuition alone, showed an acute understanding of turbulence in the twists and swirls of his painting *The Starry Night*.

---

7 Parker, "Win a Million."

Figure 0.1 *Vincent van Gogh*, The Starry Night *(1889)*.
Public domain. Image via Wikimedia Commons
(MoMA, New York).

Starry Starry Night

Those that interpret this painting often see the currents and eddies depicted as reflections of Van Gogh's turbulent mental state. He had recently admitted himself to an asylum in Saint-Rémy-de-Provence, France, following a particularly difficult period in his life and was a patient there when he created the work. Van Gogh revealed a particularly accurate understanding of turbulence through his brush strokes.

Several papers have been published over the years that analyze *portions* of Van Gogh's painting and how the swirls and eddies it contains behave in the way described by Kolmogorov's theory of turbulence, the best model we currently have. It is noteworthy that this theory was not developed until after Van Gogh's death in 1890. In September 2024 a team of French and Chinese scientists published a paper which found that in terms of position, size and luminance, the features of the *entire* painting closely adhere to Kolmogorov's theory.[8,9]

How could Van Gogh's creative sensibility possibly compose an image with such mathematical precision? This question was posed to one of the paper's authors (François Schmitt) by Bob McDonald, host of the CBC radio science show Quirks and Quarks. Schmidtt responded that rather than thinking in mathematical terms, Van Gogh "felt into it." Apparently, the felt-sense of human intuition has the capacity to bestow deep understanding, even in the realm of science.

In the following chapters we will follow Van Gogh's example by *feeling into* our own mental turbulence using the latest discoveries in neuroscience to guide our discussions. Perhaps like Van Gogh, we will gain intuitive insights on reality that have, to this point, evaded formal elucidation.

World Views and Human Values

Byron Wood, the nurse cited earlier in this introduction, is a *modernist* who values autonomy, science, and rationality over the *traditional* values of authoritarianism, order, and holiness. When modernity and tradition clash, the traditional worldview takes on the role of dominator. This is the cultural environment in which many Generation Xers, including me, were raised. It is a major factor that drives Generation X, and the generations that followed, to claim "I am spiritual, but not religious." The domination mode of the traditional worldview exhibits the kind of paternalism that catapulted me beyond modernity and into the *post-modern*, where the world is seen as a shared community for all humanity, with Mother Earth as their home. The idea that men had feelings was fanciful to my parents' generation (they were supposed to be like rocks), but the post-modern worldview gave me freedom to explore my own subjective experiences and ask questions about consciousness that would appear alien to my parents. I provide *my* answers to many such questions in these pages.

My generation, and later generations, tend to conflate traditional values with this unhealthy domination mode of operation without considering the healthy attributes the traditional worldview embodies. In reality, the traditional is the foundation on which modernity is built,

---

8 Ma et al., "Hidden Turbulence."
9 Gometz, "The Hidden Physics."

and tradition provides a safety net for those who lose their balance while exploring modern and post-modern freedoms. This book builds on the traditional, and embraces modernity and the post-modern, creating an integrated view of the milieu in which Western society operates. Each of these worldviews, and their associated values, contribute much to a healthy psyche. The aversion felt by modernists towards the Bible might be at least partially ameliorated by viewing it as an early psychology textbook. This is the perspective I take in the chapters that follow.

To those who have doubts about a science book that builds on a Judeo-Christian foundation I ask but one thing—please do not judge the traditional in the way it judged you. In the words of Mark Twain:

> An open mind leaves a chance for someone to drop a worthwhile thought in it.

Themes

The following themes are woven into the fabric of this book. In sartorial terms, they provide the longitudinal threads (the warp) into which the weft of each chapter is interwoven. Look out for these primary motifs as you make your way through these pages.

Needs: A need is a resource that is required to maintain balance in the mind-body system. These resources can be as concrete as air, water, and food, or as ethereal as beauty, autonomy, and creativity. Some of these resources are required for our survival, others are needed for us to function at the peak of our capabilities and experience rewarding lives.

Feelings: These are internal felt-sense states that emerge and fade over time. When needs are met in the present moment, we experience *rewarding* feelings like elation, enthusiasm, and serenity. When they are unmet, we experience *aversive* feelings like irritation, anxiety or boredom. Feelings arise in response to events that unfold throughout the day, but what we feel in any one situation depends on how we have been conditioned by prior life experience.

Neural Activity: Feelings are generated by the flow of electrical impulses through the neurons that comprise our central and peripheral nervous systems. We seek out the rewarding feelings and avoid the aversive ones. In situations where aversive feelings are experienced chronically or if they are unbearably intense, we might engage in behaviors or ingest substances that produce good feelings and suppress the bad. These strategies take effect by altering the flow of electricity through our nervous systems and by activating some parts of the brain and deactivating others.

Relationship: The relationships we have with ourselves and others meet a host of connection needs. Our brains are hardwired for relationship at the molecular level and when we engage with one another, our brains release and process substances that embody the unique quality of a particular relationship. If we enjoy our interactions with a person, we have *good chemistry* with them. While this term is a metaphor signifying a profitable chemical reaction between two substances, it also describes what unfolds at the biological level as neurochemicals like dopamine, oxytocin, serotonin, and the endorphins interact with neurons. Many of the substances we abuse release or mimic these same neurochemicals, empowering them to emulate the experience of *good relationship* in our lives. Without good relationships with ourselves and others (including our higher power,) we experience the pain of disconnection, and it is often this pain we seek to escape by using. As it is in relationship that most of our psychological needs are met, it forms the line that runs most prominently through the pages of this book.

What you can do now: Each chapter concludes with a contemplative or self-study activity relating to the chapter. These are gentle nudges in the direction of recovery and cannot replace the wisdom of a professional counsellor. However, they can provide nourishing fodder for your daily meditations and discussions with others. Some of these provide guidance on using affordable, available technologies. As the technological landscape changes on a continual basis, it would be unwise for me to mention specific product offerings due to the risk of obsolescence. An internet or app store search using keywords from the activity should get you what you need.

Chapters 1–5 of this book (part one) help us recognize the face of addiction; chapters 6–13 (part two) explore the tenacity of addiction's hold. The remaining chapters 14–17 (part three) provide strategies that help heal our past hurts and take care of ourselves with love. These chapters encourage us to live lives that produce rewarding feelings by meeting needs, rather than imitating those feelings using destructive behaviors and substances. I encourage you to follow this cycle of steps as you progress through the book:

1. Identify past experiences that produced rewarding feelings.
2. Explore the neurological processes underlying those feelings (gain understanding.)
3. Acknowledge that our destructive strategies are engaging the same neurological processes.
4. Do more of the things that naturally produce good feelings instead of fabricating them.
5. Accumulate the wisdom required to build *internal* resilience and contentment.

The self-compassion, curiosity and understanding these steps have brought into my life have been transformative. I hope that you are able to reap the same benefits.

Part One—*Recognition:* Identifying and Naming Addiction

1

Inching Towards the Light

"Addiction is neither a choice nor primarily a disease. It originates in a human being's desperate attempt to solve a problem: the problem of emotional pain, of overwhelming stress, of lost connection, of loss of control, of a deep discomfort with self. In short, it is a forlorn attempt to solve the problem of human pain" —GABOR MATÉ

I AM THE ALPHA and the Omega, the beginning and the end (Revelation 21:6.) With these words Christ claims oversight of the eternal order and extends promises to those who love him, those who are citizens of the New Jerusalem coming down out of heaven from God. Although this passage from the Christian New Testament is framed in temporal terms, insinuating that time has a beginning and an end, the implication is one of an eternity with Christ, a tangible revelation of God, at its center. His physical constitution, mission and purpose were foreknown from the beginning of time. The same is also said of those who follow him who are described as "chosen in Him before the foundation of the world" (Ephesians 1:4.) But what does this say of our essential nature? Are we, as human beings, eternal, without beginning and without end? Did we exist before the foundation of the world, or was our selection by God through his foreknowledge of the people we would become? Were we recruited from a swirling pool of souls (each endowed with the attributes of eternity) and infused into material bodies at some time before birth? Perhaps our consciousness arises progressively as we are woven together in our mother's womb (Psalm 139:13)? Although many faiths and traditions offer answers to these questions, the Bible remains largely silent on the matter. When the Bible speaks of eternal life in direct terms, it does so regarding the resurrection from the dead.

What Is Consciousness?

It might be suggested that prior to birth our state is identical to that after death and that a consideration of near-death experiences might shine a light on this nascent phase of existence. Near-death incidents are often accompanied by out-of-body experiences. This was the case for Tony Cicoria, an Orthopedic Surgeon from Upstate New York who suffered a lightning strike while using a public pay phone.[1] His account of the ordeal describes his perception of consciousness as perfectly normal, and yet he was able to watch people trying to resuscitate his lifeless body. During this time, his senses were situated in a location in space past which people were walking, and he could follow them with his "eyes." Experiences such as this have raised questions about the true location of consciousness. Is consciousness merely the result of neural activity? Is it really situated in the brain? Others have questioned whether such experiences are what they appear to be, or might they be the result of some trauma-induced mental process, the nature of which remains a mystery.

Our limited ability to describe consciousness in objective terms was brought home to me in the final class of a master's level course in artificial intelligence (AI.) Our professor took us on a journey through the history of western thinking on the subject, beginning with the Greeks and culminating in the present. After a lifetime studying state-of-the-art solutions to the acquisition and processing of expert knowledge, he admitted that, although these approaches had met with some success, they did not employ the same processes used by the human brain. He also admitted that despite two thousand five hundred years of philosophical thought on the matter, humanity is no closer to describing these processes. He closed the class with this quote from Ecclesiastes in the Hebrew Bible (Ecclesiastes 1:9–10 NIV):

> What has been will be again, what has been done will be done again; there is nothing new under the sun. Is there anything of

---

1. Sacks, "A bolt."

which one can say, "Look! This is something new"? It was here already, long ago; it was here before our time.

AI researchers sometimes discover that their *new* ideas regarding consciousness were actually explored by the ancients. Juxtaposing science with ancient literature may seem incongruous at first, but rejecting the wisdom of millennia merely because it appears unscientific may be discarding our best insights into the human condition. This book takes a thoroughly evidence-based approach to the subject at hand but will make frequent Biblical allusions when these shine light where science sees only dimly.

The Development of Consciousness

What we can learn from observation regarding consciousness is that once we were small and now we are grown. An analysis of early childhood development reveals that the attributes of growth extend far beyond physical dimensions. Each of us also develop morally, spiritually, and intellectually through similar stages, passing through the same general phases of physical and cognitive development. We are all human expressions of the same basic gene stock, and from this template our bones, muscles, sinews, nerves and organs take shape. Brain development is also expressed as an unfurling of genetic wisdom, developing as it does from bottom to top and right to left.[2]

Coincident with these phases of physical and cognitive development, we grow into people. We accumulate likes and dislikes, experience laughter and tears, learn words and movements. That this is both time-consuming and challenging is reasonable as we are becoming accustomed to life itself, learning what it is like to have senses and a body, discovering what control we have over our limbs and objects in the external world. We could liken this process of development to a coming into focus, a gradual revealing or a dawning from darkness into light. This transformational unfolding of sensibility has been described as Enlightenment—"insight or awakening to the true nature of reality." Rather than seeing enlightenment as a prize, the possession of which affords the owner a measure of social capital, I prefer to frame it as an ongoing and never-ending process. We are all students in this school of unfathomable complexity, and none of us can claim the title of *master*. As the years go by, we

---

2. Tierney and Nelson, "Brain Development," 2.

pass through many stages of development on this journey of progressive enlightenment, each of which corresponds with a perspective on reality or level of conscious awareness.

The goal of this book is not to promote a particular religious perspective or set of dogmas, but rather to harmonize the scientific, secular and religious tapestries that portray the addiction experience, inscribing on the resulting map a pathway leading the reader away from compulsive and destructive coping strategies. As the title suggests, spirituality is at the heart of this discussion. But what do we mean by "spirituality"? An examination of Biblical terminology is helpful in this regard.

Spirituality

Spirituality is a term derived from Judeo-Christian tradition. Wikipedia defines it as "a religious process of re-formation which aims to recover the original shape of man."[3] According to Genesis 1:27, humanity was made in the image of God. This is, in the Judeo-Christian context, our original shape, but it is a shape that each individual struggles to achieve and maintain. Regardless of your religious views, it is hard to deny that this process of re-formation plays a significant role in the cultures of the world.

Etymologically speaking, spirituality derives from the word *spirit*, a concept that runs deep throughout the pages of the Bible. In the gospel of Luke we read of the reaction shown by Christ's disciples to his unexpected post-resurrection appearance. The shocking revelation took place as two of his followers, who had just encountered the risen Christ on the road to Emmaus, came to report their experience to the disciples in Luke 24 36–39 (NASB):

> While they were telling these things, He Himself stood in their midst and said to them, "Peace be to you." But they were startled and frightened and thought that they were seeing a spirit. And He said to them, "Why are you troubled, and why do doubts arise in your hearts? "See My hands and My feet, that it is I Myself; touch Me and see, for a spirit does not have flesh and bones as you see that I have."

Of note here:

---

3. Wikipedia Contributors, "Spirituality."

1. The disciples' fear was based on their belief that Christ appeared in the form of a spirit.
2. He appeared before them with a corporeal resurrection body.
3. A spirit does not have flesh and bones; it is without substance and cannot be touched.

The disciples were fearful they were seeing a ghost, a disembodied relic of the deceased. The word translated spirit here is the Greek "pneuma," from which the English word "pneumatic" is derived. In its basic sense pneuma means *breath*. In stoic philosophy it denotes the vital spirit, soul, or creative force of a person manifested in a corporeal body.[4] Three associations between pneuma and spirit are:

1. The effects of breath can be felt but it cannot be grasped; it is unsubstantial.
2. The reality of a person's conscious life essence can be inferred from their actions and their ability to reason, but it cannot be touched or grasped.
3. Breath is vital in maintaining the life force of a person.

Spirit as a Manifestation of Consciousness

1 Corinthians 2:11 (NASB) identifies the spirit as that attribute of a person that is aware of their thoughts:

> For who among men knows the thoughts of a man except the spirit of the man which is in him? Even so the thoughts of God no one knows except the Spirit of God.

In Buddhist terms this spirit is the *observer*, the conscious entity behind the voice in our head that perceives both thought and internal sensation. We find the root word *spirit* distributed liberally throughout the English language in the word fragment "spir." When the life-force of God entered the prophets, filling them with messages of rebuke, prognostication, and encouragement, the process is referred to as *divine inspiration* (God-breathing.) The same intake of spirit-consciousness was believed by the Greeks to underly all acts of creativity, during which a subject was briefly inhabited (inspired) by a benevolent or benign nature spirit.

4. Rubarth, "Stoic Philosophy."

Conspiracy, literally interpreted, means *breathing together*. The process of breathing, including the transportation of oxygen by the blood and its utilization in energy production is known as *respiration*, and when we die we *expire* (breathe out.)

While James 2:6 (NIV) deals with far weightier issues than mere word etymology, it does throw some light on the dependence between body and spirit:

> As the body without the spirit is dead, so faith without deeds is dead.

Spirit is the intangible essence of life that contributes to humanity all aspects of sensibility, intelligence, memory, and information processing. Our spirit manifests in the process we call consciousness, without which "we" would not exist. All our defining attributes (including our character, personality, values, motivations, loves, and faith) flow through the channel of consciousness. When unconscious, we enter a state that resembles the sleep of death and become a body that, while it may still be living, is devoid of *this* kind of spirit. When God chose us before the foundation of the world, He based this choice on the "us" expressed in these spiritual attributes.

Spirituality in the Ecclesiastical Sense

Historically, the word *spirituality* has been used by the Christian Church to describe structure, order, and influence in the Ecclesiastical sense.[5] In so doing, it associates spirit (and by extension *spirituality*) with the manifestation of Christian authority and orthodoxy. This book uses the word in its more general sense as anything concerning conscious perception, which may or may not relate to a manifestation of the Godhead. However, I believe it is impossible to examine our own spirituality without reference to the one great spirit that we, as humans, manifest. A goal in these pages is to draw a clear picture of the role the higher power/spirit plays in the addict's recovery and how that power gives structure to the self.

---

5. DeYoung, "Two Cheers."

The Body/Spirit Dichotomy

Thus far I have presented spirit as something distinct from the body, without which the body would cease to function. This distinction between the body and mind is not clear-cut, though. Our consciousness *is* affected by what happens to our body. We drink too much coffee and are unable to fall asleep. We are scheduled for an operation and are rendered senseless by the general anesthetic. We ingest a psychedelic substance and enter an altered state of being. Whether out-of-body experiences provide evidence for the existence of a disembodied "me" or not, we must admit that our general participation in the here and now is neurochemical in nature. For this reason, our discussion of "spirit" will focus on the sensibility of the physical body.

God is spirit (John 4:24) and it seems implausible that the one who existed before the universe came into being would have a physical body. With this in mind, we can consider the consciousness manifested in humanity as a kind of material/electro-chemical metaphor of the ultimate spiritual reality. Could this be what is meant by "Then God said, 'Let us make mankind in our image'" (Genesis 1:26 NIV)

How, then, does spirit manifest in our physical bodies? Our nervous system operates by passing messages between nerve cells (neurons) using neurotransmitters. These molecules act as keys that activate receptors in the receiving neuron, thus triggering an electric current, or potential. Some neurotransmitters increase the flow of electricity into the receiving neuron, while others reduce it. This process is influenced by substances that enter our bodies from the outside, allowing these chemicals to bring about changes in our state of consciousness. Through caffeine's activity, the brain's adenosine receptor system is inhibited, bringing about a state of mild agitation. The general anesthetic activates the GABA or deactivates the Glutamate receptor systems, causing a reduction in electrical activity to a point below the threshold required for conscious awareness. The psychedelic substance activates serotonin receptors, resulting in altered perceptions of thought, space and time.

Our Introduction to Consciousness

The freshly minted human is unaware of any of this and experiences these chemically mediated processes as *consciousness*. The infant at birth is a sensorimotor organism possessing neither linguistic ability nor an appreciation of time. They have no sense of the subjective dimension but exist in a sensoriphysical space. As time progresses, we learn the distinction between our bodies and the physical world. Still later we learn to differentiate between our own emotions and those of our companions, including those of our mother. These mark the first two stages of personal development as defined by Margaret Mahler, Melanie Klein and others.[6] Stage three is the first to allow for introspection, the ability to sense and identify internal physical and emotional states.

I have vivid memories of two emotional discoveries at this third stage of life, memories that were given weight by the novelty of the experience. While being carried by my mother, I noticed a pleasing inner glow as I laid my head on her shoulder. I was familiar with many emotions common to a child, all of which seemed unremarkable, but this one caught my attention. Even at this young age my musings on the strangeness of existence, of life, consciousness, and the material world had become part of my everyday experience. What were the origins of these realities, and what was their true nature? Societies throughout time, who shared this childlike worldview, asked the same questions and created cultural narratives to explain the human experience. The First Nations of Haida Gwaii possess a rich mythology with fantastic accounts of the origin of all things. The Mesopotamians of the Euphrates and Tigris basin had their own creation stories and the Egyptians still others. These mythologies provide culture-specific answers to the how and why of human existence but fail to elaborate on the mechanisms of our subjective experience.

As a child in my mother's arms, there was nothing I could draw on to account for this sensation. What was the objective difference between a pleasant and unpleasant sensation? I was grasping at air, trying to hold on to the unsubstantial, attempting to nail spirit to the wall. It was a losing proposition. Little did I know that philosophers and scientists, from time immemorial, had struggled to answer the same questions with little more success than I could muster. To twenty first century people, consciousness is still a mysterious capacity unexplained by even the most advanced research techniques and models of neural function, so what hope did I have of solving this riddle?

---

6. Wilber, *A Brief History*, 147–48

Early Relational Behaviors

My relationships with other conscious beings also spoke to my naïveté. Being raised in an agricultural environment, I was accustomed to speaking with creatures that lacked linguistic abilities. Horses, cows, and chickens were my companions as were the ever-present gaggle of family pets. I even remember discussing the possibility of life on Mars with a Palomino pony called William! Add to that the prime-time TV spots describing the benefits of talking to our plants, and my imagination really took flight. I talked to African violets, dahlias, and horse chestnut trees. If they were big enough to hug, I would hug them! What I was reconstructing were the beliefs held by ancient cultures—belief in the spirits within plants, the earth and water. I even had a few pet rocks! Aspects of this thinking were persistent. Even at eighteen years old I believed my Austin Mini possessed conscious awareness. It did, after all, have electrons flowing through its frame. It came alive at the turn of a key and performed tasks as directed; why would it not be conscious? It all seems so illogical now, but such is the reality/perspective of the agrarian worldview. (While those with the rational worldview of twenty first century Western culture may scoff at these ideas, there is a growing body of scientific evidence that plants have a form of awareness with at least some of the features possessed by mammals.)

Peak Experiences

A year or so after the moment of transcendent connection with my mother, I was playing on the floor of my bedroom when my attention fell on a corner of the bedspread as it swept over the edge of the bed and lay diagonally across the carpet. The sun was beaming through the window warming the environment and bringing out the rich greens, blues, and highlights of the wool fabric. I was transfixed by the sensation of warmth and coziness instilled by the sight. The sensation was similar to the glow brought on by my mother's nurturing touch, but it was even more intense. Rather than accepting this sensation in a non-judgmental way, my reaction was one of confusion and fear. Having had two experiences with similar qualities, however, I was convinced that the first was not an anomaly but an aspect of conscious awareness that might be reproducible. It was certainly desirable to be in such a state, but I had no idea how one might create the conditions for its expression. These events could be described, using the language of Abraham Maslow, as *Peak Experiences*.[7]

Dopamine—an Introduction

While the collective musings of humanity may have missed the mark in framing these pictures of pain and ecstasy, we do understand the neurochemical prerequisites for such states, and they have much to do with a tiny signaling molecule called dopamine. It is not that dopamine itself creates such transcendent experiences, but it does drive the propagation of electricity through the incentive and reward pathway, part of the primitive brain that governs and regulates human motivations and their behavioral responses. As the movement of skillful fingers brings a stringed instrument to life, so the flow of dopamine propagates resonant frequencies through the pathways that drive us to action and reward us for so doing. We can measure the chemical and electrical realities behind this process, but without an understanding of what constitutes consciousness, we lack the language to describe anything beyond the realm of electro-chemistry and physiology. What we do know is that all addictions manifest within the incentive and reward pathway (or limbic loop) so it is here that much of our discussion will be focused.[8]

One thing to note is that scientific knowledge evolves over time as new discoveries are made. In the period from 1990 to 2010 it was believed that dopamine was wholly responsible for the feel-good quality of *liking* something. Reality, however, is a bit more nuanced. While the experience of euphoria is still ascribed to the same brain region (the nucleus accumbens,) it is now thought to involve the downstream activity of opioids on hedonic hotspots in the nucleus accumbens shell[9] rather than being the responsibility of dopamine alone. However, as this opioid activity is part of a feedback loop that involves other brain regions and includes dopamine in its lines of communication, I will be presenting dopamine as the source of both incentive and reward for the sake of simplicity. What the evidence still shows is that Dopamine drives both our cravings and addictions.

---

7. Maslow, "Religions, Values", 4
8. Halber, "Motivation."
9. Peciña and Berridge, "Hedonic."

The limbic system is a set of structures, located directly behind the eyes, that process emotions and memory. It regulates autonomic (subconscious) and endocrine (hormonal) function in response to emotional stimuli and is involved in reinforcing behavior. Behaviors that we partake in are generally reinforced because we feel good in response, and it is this feature of felt-sense awareness that leads to addiction. We ingest a substance or partake in an activity that produces a sense of well-being, relaxes us, eases our stresses, worries, and fears, and we want to reproduce that state of being at-will. This state, usually described as being high (of elevated mood), is generally related to dopaminergic activity in the nucleus accumbens. The enjoyment that flows from such diverse stimuli as sensory incitement, achievement, creativity, and social connection is an endowment of the limbic loop. As our conscious experience of life is created by the propagation of electricity throughout our brain and body, an understanding of this process helps to pinpoint key attributes of our underlying dysfunctions.

Meeting Needs in All the Wrong Places

Addictions are our response to unmet needs, but they are dysfunctional responses. They need not involve the abuse of substances, but they always involve behaviors that are harmful to ourselves and/or others. Despite causing harm, the addicted individual will pursue these behaviors compulsively and without any apparent resolution. We may be addicted to gambling, relationships, touch, the regard of others, control, prestige, love, shopping, perceived freedoms, money, power, music, religious activity, and many other elements that are sometimes okay in and of themselves, but become impediments in how we relate to them. We are designed to have these needs met, but only in ways that are life-serving. In each case the process of addiction involves the same neural circuits, whether they are driven by signaling chemicals produced inside the body (endogenous) or by introducing analogues of those chemicals from the outside (exogenous.) Chances are we are all addicted to something, a fact that ought to moderate how we perceive the vicissitudes of others. Ultimately there is only one addiction—the addiction to dopamine—and this is the common thread that links all behavioral and chemical dependencies.

Compassionate Curiosity

Dr. Gabor Maté's book *In the Realm of Hungry Ghosts: Close Encounters with Addiction* contains a chapter entitled "The power of compassionate curiosity."[10] He begins this chapter by describing how his addiction of choice (compulsively buying compact disks) had dogged him throughout the writing of the book and how the associated binging and lying left him feeling shamed and hollow. He saw his greatest triumph in the *pursuit* of recovery rather than in its culmination, a result he saw happening in the distant future. All of us do things we would rather not do, and Dr. Maté sees our willingness to express a *compassionate curiosity* towards ourselves as the best starting point in addressing these foibles. The most pertinent question to ask ourselves in this regard is the following: why do I do what I do? This book is written in the spirit of compassionate curiosity, not just for me personally but also for all those who seek the best for themselves and others. Curiosity that is compassionate is not impeded by self-judgement or self-loathing. It involves a clear-eyed acceptance of the observed facts at face value and the recognition of denial, wherever it rears its ugly head. Maté developed this concept into a counselling modality known as Compassionate Inquiry, the details of which are outlined in his book *The Myth of Normal*.[11]

My curiosity began with accepting that I do things because they make me feel better and lessen the underlying sense of unease and/or emptiness. This observation indicates that *feelings* are the main driver of both life-serving and destructive behaviors and that answering the following question is of prime importance: If I do things that I would rather not do because they make me feel better, what is the mechanism underlying the unease from which I flee and the relief that I seek?

I could have undertaken a literature search and presented my findings, but this would have added little to the value of this book other than in presenting a digest of existing materials. What I offer here is a description of the felt-sense attributes of human consciousness from a phenomenological perspective.

---

10. Maté, *In the Realm*, 329.
11. Maté and Maté, *The Myth*, 410.

Phenomenology

Phenomenology is a branch of philosophy introduced by Edmund Husserl[12] that concerns the systematic reflection on and study of the structures of consciousness and the phenomena (feelings, perceptions and judgements) that appear in acts of consciousness. As we are all experts in being conscious, this approach offers each of us a powerful research tool with which to exercise compassionate curiosity. The phenomena (felt-sense experiences) that I will be exploring relate to the feelings and states of being that move us to action and cause us to do what we would rather not do. My goal is to help every reader gain an intimate understanding of their motivations by framing their own lived experience as an intentional research and exploration process.

Our first step in exploring the phenomena expressed by the mind, body, and spirit is a consideration of the human nervous system.

Thematic Takeaways

Feelings: Peak experiences involve profound feelings of joy, awe or transcendence. We can use substances and behaviors to artificially manufacture such states in unsustainable ways, but not without harming ourselves or others in some way.

Needs: Addictions come into being when we live in a state of chronic emotional pain. Addictions mask the pain of unmet need by creating a perception of wellbeing that soothes, excites or distracts from the pain we experience in the present moment.

Neural Activities: Our experience of consciousness manifests in the electrical currents that flow through the neurons of our central nervous system. This flow is generated by the passage of neurotransmitters between one neuron and another. We can transform our state of consciousness from a disagreeable one to a pleasing one by ingesting a substance or engaging in a behavior that alters this transmission process.

Relationship: The feelings I experienced in my mother's arms were sparked by the love I had for her, the trust I had in her and the attachment I experienced with her as a caregiver. These higher-level mental functions were expressed sensorially by the release of oxytocin,

---

12. Husserl, *Ideas.*

serotonin, dopamine and other related neurochemicals. This mechanistic perspective in no way detracts from the sanctity of our relationships but brings clarity to the process that motivates the establishment and maintenance of these bonds.

What you can do now: Begin familiarizing yourself with the many human needs identified by proponents of Non-Violent Communication at baynvc.org/list-of-needs/. While this book is not a training manual for NVC, it draws on the concepts of feelings and needs as fundamental drivers of the addiction process.

2

Fearfully and Wonderfully Made

> "For you created my inmost being; you knit me together in my mother's womb. I praise you because I am fearfully and wonderfully made; your works are wonderful, I know that full well." —Psalm 139:13–14 (NIV)

To begin a discussion of how addiction manifests in a person we need to know how "a person" operates. In the following pages I attempt to lay the foundation for this understanding, although the discussion is necessarily technical. Addiction is realized through various neural circuits and the chemicals that innervate them. Taking a layman's approach, I limit this discussion to systems you may already be familiar with as they are discussed regularly on TV, in the press and on social media. Once we have this knowledge, discussing these systems becomes much easier. Internalizing the knowledge, however, can be challenging. The only reason I explore this in some depth here is to lay the foundation for a deeper understanding of the chapters that follow. For the sake of brevity, I have also kept descriptions of the terms used brief, which gives this chapter a rapid-fire vibe that some might find overwhelming. If you find that reading the following material is an impediment to continuing with this book, you could skip to chapter 3 and look up the meaning of any terms you do not understand as they arise.

The Central, and Peripheral Nervous Systems

We are all familiar with how our mind interacts with our body to effect movement. We create an intention to move our arm, and it responds. This is a function of our central nervous system (CNS) that we learn in the very earliest stages of life. The brain and spinal column comprise the CNS, which concerns itself with thought, hormone activation, action, and sensation. The other branch of our electrical infrastructure is the peripheral nervous system (PNS,) a major component of which is the autonomic nervous system (ANS.) The ANS concerns itself with the housekeeping tasks of pumping blood, digesting food, breathing, and other tasks that can be successfully undertaken without our conscious intervention. Without the ANS we would have to pump blood through our veins, digest our food, and breathe consciously! Our conscious minds would be so busy managing all that activity that there would be little capacity left to manage our higher-order functions. The ANS is the greatest helper imaginable by doing most of the functional heavy lifting while we get to concentrate on more engaging activities.

The Breath of Life

We have partial control over some of our autonomic functions (for example, we can hold our breath, and we can choose to hyperventilate,) but the ANS ensures that our breathing continues unabated while we are asleep and when we are otherwise engaged. The autonomic management of breathing is interrupted when certain drugs enter our bodies. An almost universal feature of general anesthetics is the suppression of the breathing reflex. When under a general anesthetic, a patient needs to be sustained using a respirator, a machine that breathes for them.[1] Respiratory suppression is also a side effect of substances produced by the opium poppy (*opioids*) so ingesting high doses of opioids outside of a medical setting is dangerous. Respiratory suppression is the primary symptom of opioid overdose. This breathing process has continued, in people and animals, since creation. As Genesis 2:7 (NASB) recounts:

> Then the Lord God formed man of dust from the ground and breathed into his nostrils the breath of life; and man became a living being.

---

1. Whitlock, "Discover When."

Concerning the animals, Psalm 109:24 (NASB) adds:

> You hide Your face, they are dismayed; You take away their spirit [lit. *breath*], they expire [lit. *breathe out*] and return to their dust.

When a person abuses opioid drugs, the breath (given to them by God) may be taken away.

The Sympathetic and Parasympathetic Nervous Systems

The autonomic nervous system is divided into the sympathetic and parasympathetic nervous systems. The sympathetic nervous system governs our fight or flight response. When we feel threatened, our brain tells the adrenal glands (located above the kidneys) to release adrenaline (a messaging molecule similar to dopamine) into our blood stream. This initiates a full-body response to prepare for maximum physical strength, speed and endurance. The parasympathetic nervous system generally has the opposite effect (relaxing rather than stimulating,) but the two work in a complimentary fashion. They can both be highly active at the same time; in which case the body experiences a combination of effects that may cancel each other out. To illustrate the connection between messenger release and a "felt sense" or conscious perception within our body, let's explore the adrenaline response.

Adrenaline

Adrenaline has many effects throughout the body. Many of these involve the transport of oxygen from our lungs to our muscles. These include the dilation of our airways—the nose and the bronchioles in our lungs. Our pupils dilate, increasing the amount of light reaching the retina, improving vision in low-light conditions. The arteries that carry blood to our locomotory muscles expand to increase the flow of blood, and those in our extremities constrict to move blood into those muscles involved in action.[2]

Many of these effects are palpable, although we might not notice them at the time due to the stresses of the situation. If, however, we introduce an artificial source of adrenaline into our bodies, we are more likely to perceive the change. Such is the case with the administration

---

2. Cleveland Clinic, "Epinephrine."

of an EpiPen to counteract the symptoms of anaphylactic shock. Anaphylaxis is an allergic reaction which can cause inflammation in the bronchiole linings of the lungs, making breathing difficult or impossible. In severe cases this can be fatal, but the bronchiole-dilating action of adrenaline can be a lifesaver. EpiPens, as their name suggests, deliver a fast-acting dose of epinephrine.

> Due to a patent and terminology conflict, originating in 1901, adrenaline is referred to as epinephrine in North America. The drug company Parke-Davis co-opted the traditional Chinese medicinal herb Ephedra to produce a decongestant they named Adrenalin [adrenaline without the final 'e']. Adrenaline was renamed epinephrine to avoid any confusion this might cause.[3]

I want to focus on the physical sensation of adrenaline as it moves through the body from the injection site. Some have described this as being like fear itself, which is not surprising as the same messenger is at work in both cases. Our perceptions of this condition might include a creeping feeling or a sense of anxiety. Other symptoms, such as a dry mouth, racing heart, goose bumps and sweating might also be present. While these are merely sensory phenomena, we perceive them as aspects of conscious awareness and ascribe meaning to them based on our biases and prior life experience. What the recipient of an epinephrine/adrenaline injection feels is their cell metabolism changing as the adrenaline binds with receptor sites on their cell membranes. The feeling has a unique quality that we could classify as an emotion. If you spend time tuning in to your body's reaction to being threatened, you will be enhancing your capacity for self-awareness.

One thing to notice in this situation is that the airways in your nose open as the blood vessels that surround them constrict. Depending on how blocked your nasal passages are at the time, you might feel or hear a click as the airways open. This is a response commonly exploited by the pharmaceutical industry. The constriction of nasal blood vessels, allowing you to breathe freely when you are suffering from a cold, is effected using drugs that mimic adrenaline. Were you ever amazed by how quickly and effectively a decongestant spray restored your ability to breathe through your nose? This effect is the result of adrenaline receptor

---

3. Wikipedia Contributors, "Parke-Davis," para. 14.

function. In addition to their use in sprays, decongestants are also used in pills, cough syrups, and liquid cold remedies. Two of the most common drugs of this class are pseudoephedrine and phenylephrine. Phenylephrine is also administered by optometrists to dilate the patient's pupils, so the inside of the eye is easier to observe. As mentioned earlier, pupil dilation is another aspect of the adrenaline response. Both pseudoephedrine and phenylephrine bind with the alpha adrenaline receptor to bring about the pupil dilation effect.[4]

The Multifunctional Nature of Messaging Molecules

Most signaling molecules are multifunctional—they have one function in the body and another in the brain. Adrenaline is used in the body to prepare it for maximal performance, but it is also active in the brain during the propagation of electrical signals. When our blood is teaming with adrenaline molecules, they are also being carried through the blood vessels in our brains. Imagine if the adrenaline in our blood was able to enter the fluids infusing the grey and white matter of our brains. If that were the case, it would interfere with adrenaline-based signaling in the brain. To prevent this, there is a selectively permeable membrane between the blood vessels and the brain tissue itself. This blood-brain-barrier (BBB) prevents messaging molecules that perform one function in the body and another in the brain from crossing between the two domains. The BBB is permeable, however, to a wide range of drugs which can, and do, alter the brain's operation.

Another example of compounds that illustrate this selective behavior, and one that may seem familiar to you, involves the distinction between drowsy and non-drowsy antihistamines. Histamines activate an immune response that helps the body to handle irritants such as pollen, wasp stings and physical trauma. Antihistamines are drugs that circumvent the normal action of histamines and are used to reduce swelling, itching and pain. However, histamines (in a similar way to adrenaline) are also signaling molecules in the brain. The functions they affect include sleep cycles and motion sensing. Diphenhydramine (Benadryl) is an antihistamine that is used both topically and orally, but as it readily crosses the BBB it plays havoc with the brain's ability to maintain a wakeful state. For this reason, Benadryl is sometimes used in over-the-counter

---

4. Ostrin and Glasser, "The effects of Phenylephrine."

sleep aids. It is also the active ingredient in Gravol (Dramamine.) Gravol relieves the symptoms of motion sickness by interfering with the signals passing between the inner ear (which detects motion) and the brain. It is likely that if Benadryl makes you drowsy Gravol will too, even though the latter also contains a stimulant (Theophylline) designed to counteract the drowsiness. Non-drowsy antihistamines cannot cross the BBB so they only act on histamine receptors in the body and not the brain.

Benadryl and Gravol both have abuse potential due to their blocking action on acetylcholine receptors. As acetylcholine is the messenger that feeds sensory input between the body and the brain, blocking its action dissociates the mind from the body. This provides a means of escape, for some, from their troubles and causes them to enter a hallucinogenic state.[5] However, for most people, including me, this is not an enticing prospect.

Paradoxical Drug Effects

Different people react to the same chemical signal in different ways. Some experience little sleepiness after ingesting Gravol, while others are made drowsy by non-drowsy antihistamines. This variability in response takes place with all classes of drugs and reflects differences in the way individual bodies process them. In extreme cases this can result in paradoxical reactions where a drug has the opposite effect in one individual to that in the population at large. People who receive a sedative to calm them down prior to or during surgery sometimes experience extreme levels of anxiety during recovery, which might include a racing heart. Opioid pain killers make most people drowsy, but they can also make sleep almost impossible for others. This paradox is usually a result of adaptations in the nervous system to a sedating drug and manifests during withdrawal. Withdrawal might occur in some individuals as the result of chronic sedative use, while in others a single dose might result in temporary bouts of insomnia. As a case in point, after becoming habituated to the use of opioid pain killers, Michael Jackson found sleep so difficult to attain that his doctor rendered him unconscious every night using the general anesthetic Propofol. The insomnia he experienced was the result of his chronic use of the pain killers that normally promote

---

5. Biounity, "Dimenhydrinate."

sleepiness.[6] Stimulants that are used as diet aids may likewise *increase* the appetite of some people.

Neural Receptors

In the previous chapter I introduced the idea of the 'felt sense,' a subjective awareness of a mental and physical state of being (a phenomenon) that can be explored consciously. A key to understanding the mechanisms underlying sensed state (and, therefore, the machinery of conscious awareness, mood, desire, reward, and satisfaction) is the *receptor*. Our bodies are peppered with them, and they allow messages to pass from outside a cell by signaling across the cell membrane.

When we make it our intention to move our arm, signals are propagated along the length of the motor neurons that connect our brain to the muscles that effect the motion. As I type this, I am exercising a more focused process—fine motor control. In both cases the intention and the signal to action involve the transmission of electrical currents between neurons. To accomplish this, chemical messengers are released from the upstream neuron and excite a current in the receiving neuron. This process relies on a transmitting impulse that releases neurotransmitters into the synaptic cleft—the space between the two neurons. The surface of the receiving cell contains many receptors. These are tiny molecular switches which, when bound by a signaling molecule, cause charged ions to flow across the cell membrane from the fluids surrounding the neuron. Once they enter the cell, they influence the voltage within. If the charge of the ions is positive, they promote signaling; if it is negative, they suppress it.

In our motor control example, the signal is propagated by the neurotransmitter *acetylcholine*. When an acetylcholine molecule is released by the upstream neuron, it may bind with a nicotine receptor on the surface of the receiving cell. If this happens, it opens a calcium ion channel, a tiny pore in the cell membrane through which only calcium ions can pass. Calcium ions are positively charged; so as more ion channels open and more calcium flows across the cell membrane, the local voltage inside the cell rises. When it reaches minus forty millivolts, a condition known as an action potential arises. This initiates a chain reaction, causing the current to flow along the length of the neuron and effect a muscle contraction.

---

6. ABC, "Murray trial," para. 2.

Ionotropic and Metabotropic Receptors

An important distinction to make here is between receptors that contain ion channels (ionotropic) and those that do not (metabotropic.) If manipulating the voltage within a nerve cell were the only action a receptor could take, they would not be useful in promoting any other type of cellular activity. Receptors that operate without the use of an ion channel usually employ a second molecule (or second messenger,) which resides on the interior side of the cell membrane. An example of this is illustrated by the muscarinic acetylcholine receptor commonly found in the PNS. When an acetylcholine molecule binds with the muscarinic receptor complex on the outside of the cell, it changes the shape of the receptor. This unlocks the second messenger molecule cAMP on the inside of the cell membrane and sets it free into the cytoplasm of the cell. The cAMP molecule may then initiate a cascade of reactions that lead to the desired result, which may be a muscle contraction, for example. In nerve cells the cAMP molecule could also attach to ion channels on the inside of the cell membrane and initiate an action potential, or initiate a biochemical cascade.[7] This process is slower than the fast-signaling approach used by the nicotine receptor and is utilized to initiate changes in state such as the relaxation of smooth muscle in the walls of blood vessels, the transmission of flavors from the tongue to the brain, and the modulation of mood.

The Allure of Food

Due to its limbic effects, some individuals are susceptible to the addictive properties of food. Those foodies among you will relate to the method by which the umami (savory) flavor is detected and transmitted from the tongue to the brain. The receptors responsible for this process are metabotropic receptors which are activated by the neurotransmitter *glutamate*. Glutamate occurs in elevated levels in savory foods such as meat, fish and cheese, but the levels can be increased artificially by adding monosodium glutamate to the food. Glutamate is the most common neurotransmitter in the brain but, as with adrenaline, the BBB protects the brain from being inundated with glutamate that has been ingested in the form of food. The monosodium salt of glutamate is used as a flavor

---

7. Roth, "Molecular Pharmacology."

enhancer because it is more stable over extended periods of time than glutamate alone and can withstand the cooking process.[8]

When you take a mouth full of savory food, the glutamate molecules in the food bind with the glutamate receptors in your tongue. The levels of glutamate in the food are directly related to the number of glutamate receptors bound and, therefore, the number and strength of the action potentials firing between the tongue and the brain. The "yumminess" of the food is dictated by the total number of action potentials. This is where salt comes in handy as a further flavor enhancer. Because the firing of an action potential often relies on the transfer of sodium ions across the cell membrane, the higher the concentration of sodium ions outside the cell, the more are pooled around the neuron ready to fire up the flavor sensation. Salt is an excellent source of sodium ions, so it adds to the intensity of the message transmitted. This perception of delicious flavor is a function of the limbic system's incentive and reward pathway and helps motivate us to ingest the sustenance our bodies need.

While we may eat because our body needs sustenance, we also do it because we enjoy the flavor of the food. The incentive to eat is in the anticipation of the food's flavor, which is the reward we get for chewing (in this case) on glutamate rich foods. This flavor sensation is another type of phenomenon to watch for in your process of compassionate curiosity. In his book *Recovery: Freedom from Our Addictions*[9] Russel Brand describes how all his addictive behaviors were rooted in his childhood obsession with sugar. The sweet taste of sugar is also detected and perceived in a comparable way to that of savory foods, but sweetness involves the T_1R_2 and T_1R_3 receptors rather than glutamate receptors. We experience the craving for something sweet because the incentive side of the limbic loop drives us to revel in the delicious flavors that sweet foods elicit in our *nucleus accumbens*.

Second-Level Flavor Enhancers

Returning to our umami discussion, food scientists know that glutamate is a good flavor enhancer, but this is where they get creative. Their goal is to maximize the allure of a product and ship more units. They accomplish this by adding chemicals that enhance the activity of

---

8. Jiang, "Two Sides."
9. Brand, *Recovery*, 22.

the umami receptors. This results in the opening of more ion channels and creates an even more powerful perception of flavor. Next time you are eating a bag of chips (crisps) check the list of ingredients. In North America they usually include monosodium glutamate (MSG,) disodium inosinate and disodium guanylate. In Europe, the last two may be identified as ribonucleotide flavor enhancers. These chemicals enhance the operation of the umami receptors, increasing neurological signaling and intensifying the enjoyment of the eater.[10] Because these compounds are expensive to manufacture, they are rarely used on their own unless the manufacturer wants to promote their food as "No MSG." Typically, a large boost in flavor can be achieved by including higher levels of MSG and lower levels of inosinate and guanylate. It is through these neuro-chemical processes that the felt-sense of umami is experienced. This approach blurs the line between food and drug, especially as it relates to limbic activity, but in this case the "drug" acts locally on the tongue rather than globally on the CNS.

Agonists and Antagonists

Receptors are complex structures composed of tightly-folded proteins. Many of them consist of seven domains (or folds,) which loop from the outside of the cell membrane through the membrane to the inside and then out again. This presents a complex exterior surface that is subject to interactions with many different compounds. This is a key feature of drug design as many classes of drugs act simply through their ability to bind in diverse ways with their target receptor. Some drugs activate the receptor directly (agonists); others deactivate it (antagonists.) Still others, such as the benzodiazepines (Valium, Ativan, etc.) modulate the affinity of the receptor for its natural signaling molecule (allosteric modulators.) These perform a function like that of an agonist without introducing the same level of overdose risk. Allosteric modulation is also the mechanism by which inosinate and guanylate operate on the umami receptor.

Antihistamines are histamine *antagonists*. They block the action of the histamine receptor and prevent histamines from having any effect on the function of the host cell. As antihistamine levels rise, more receptors are blocked, and the effect becomes progressively more noticeable. While the discussion of antihistamines may seem peripheral

10. Kurihara, "Umami."

to the subject at hand, we should note that the antihistamine *promethazine* is a commonly abused party drug. I will have more to say about promethazine this later in this book.

In this chapter I introduced the biochemical processes underlying many felt-sense states. In the following chapter I will focus on how compassionate curiosity helps us become aware of these visceral and mental processes. With an awareness of the physicality of consciousness, we can detach our sense of self from how we feel in the moment (it is just chemistry) and view our feelings as acts of consciousness over which we have some control.

Thematic Takeaways

Feelings: The activities of signaling molecules within the body are palpable. The quality of these feelings can be explored consciously to determine which molecules are active, based on the felt-sense they evoke. Knowing which molecules are active helps us to objectify those states and disidentify ourselves from them. Seeing ourselves as independent agents in this process helps us to let go of our compulsions. With the knowledge that it is dopamine that makes us feel good and drives us to want more of what we are already enjoying, we can identify our craving as just dopamine doing its thing. This perspective makes the craving less of an imperative and reveals it as a mere function of body chemistry.

Neural Activities: Receptors are the activation switches for various bodily functions. Signaling molecules bind with these receptors to trigger a discrete response. The Medulla Oblongata (located just above the spinal cord) contains many mu opioid receptors that bind with opioid drugs. When bound, they suppress both heart rate and respiration. This is the most common symptom of opioid overdose and can starve the brain of the oxygen it needs to function.

What you can do now: Begin exploring the felt-sense states evoked by various life circumstances and learn to identify them as electrical phenomena. The goal is to see them as expressions of lower-order neural activity rather than a reflection of who you are. Of special note are occasions when you long to escape bad feelings or amplify good ones.

3

Self Awareness

"It's a lifesaver—adrenaline. I think I have an adrenaline addiction, no question about that." —Tom Waites

THIS CHAPTER EXPLORES THE effects of various pharmaceuticals on the conscious mind. It is not a critique of the pharmaceutical industry, but rather a dispassionate consideration of my own felt-sense observations. Treatment with pharmaceuticals involves an assessment of the costs and benefits relating to their use. Additionally, a particular treatment may have no benefit for one individual but have life-changing benefits for another and as such this is only *my* account. We must also remember that these substances are effective only when used as prescribed and exceeding the recommended dose may cause harm.

Inhibiting the Immune Response and Reducing Inflammation

At the age of sixteen I was diagnosed with chronic sinus inflammation and the buildup of fluid in my sinuses. My doctor, using the best knowledge available at the time, prescribed the drug Actifed. Actifed contains two active ingredients: pseudoephedrine (a decongestant) and triprolidine (a sedating antihistamine).[1] The goal of using this drug was to reduce sinus inflammation and the production of fluids. It was unclear to me whether there was any positive effect on my condition. What was evident, as soon as I started treatment, were the side effects

---

1. Medicines.org, "Multi-Action ACTIFED."

experienced by my brain. Pseudoephedrine is the same compound sold as Adrenalin under the patent obtained by Parke-Davis mentioned earlier. It is an adrenaline agonist; it activates adrenaline receptors. Triprolidine is a histamine antagonist; it deactivates histamine receptors. Both compounds cross the blood-brain barrier (BBB) and affect histamine and adrenaline signaling in the brain. The combined result was the long-term (several months) partial activation of my fight or flight response, combined with a debilitating histamine related drowsiness. Both sensations were quite unpleasant.

One thing I learned from this was how to identify the felt-sense experience of histamine antagonism. This became useful later in my development of self-awareness and my understanding of drug/receptor function, enabling me to identify which receptors were being bound by a particular drug from the felt-sense they evoked. This awareness also helped me pinpoint what was going on in my body and brain, both physiologically and emotionally, during an entire range of life experiences (experiences that produce endogenous chemical responses.) At a low point in my life my doctor prescribed the anti-depressant Serzone.[2] This had little effect on my mood, but it did make me very drowsy during the day. Shortly after discontinuing my treatment with Serzone, my doctor prescribed a course of amitriptyline.[3] It produced side effects like those of Serzone. It was at that point that I noticed the similarity between the drowsy sensation produced by triprolidine, Serzone, and amitriptyline. After further research, I identified all three of these drugs as histamine antagonists. Each of them has a primary function in the treatment of ill health, but because they interact with multiple receptor types, they also affect signaling pathways that have nothing to do with their intended function. This is one of the primary reasons why drugs have side effects. It is hard to develop a drug that selectively binds with one kind of receptor without affecting others.

A Phenomenological Approach to Drug Research

Having identified this sensation as a feature of histamine antagonism, I started searching my memory for other times I had been in a similar state to see if that might be a result of another type of treatment. I was able to

---

2. Wikipedia, "Nefazodone."
3. Ro.co, "Amitriptyline," para. 16.

recall the felt-sense arising from several pairs of drugs and compare them for similarities. Using this technique, I realized that the antibiotic tetracycline had produced a similar felt-sense to these antihistamines. Further reading showed that, although tetracycline is not an antihistamine (it is a metalloproteinase inhibitor,) it inhibits the *release* of histamines.[4] This makes an important point—similar effects result from blocking a receptor and preventing the signaling molecule from being released in the first place. The result of blocking the histamine receptor (histamine antagonism) and preventing the release of histamines is to suppress histamine function, with comparable results. I concluded that, in the absence of other stimuli, any dizzy, spaced-out, or cozy sensation stemmed from the psycho-active effects of some medication or its endogenous counterpart.

You may wonder why I am paying so much attention to the effects of adrenaline and antihistamines? Sadly, exogenous sedating antihistamines and endogenous adrenaline both have abuse potential, making them relevant to a consideration of addiction. (You are probably familiar with the term "adrenaline junkie")? Understanding their modes of action also lays the groundwork for an understanding of more commonly abused drugs, such as the amphetamines and the opioids.

King Solomon's Phenomenological Exploration of Value and Pleasure

King Solomon embarked on a similar phenomenology research project to the one I am presenting here. His focus was on the operation of the nucleus accumbens—that part of the brain that creates the experiences of motivation, pleasure, and the rewarding sensations that accompany creativity and the appreciation of our accomplishments. In Ecclesiastes 2:1–3 (NASB) he lays out his approach:

> I said to myself, "Come now, I will test you with pleasure. So enjoy yourself." And behold, it too was futility. I said of laughter, "It is madness," and of pleasure, "What does it accomplish?" I explored with my mind *how* to stimulate my body with wine while my mind was guiding *me* wisely, and how to take hold of folly, until I could see what good there is for the sons of men to do under heaven the few years of their lives.

Solomon's curiosity was not just in an exploration of pleasure, but also of folly. He explored the felt-sense effects of alcohol to see what was

---

4. Runsewe, "The Inhibitory Effects,", para. 6.

gained from the pleasure it evokes and delved into the shadow side of the human condition to determine why humans act in self-destructive ways. In both cases he saw their rewards as meaningless. (We will explore how the shadow side of our nature contributes to addiction in chapter 15.) For a pop culture perspective on the harm arising from drug use, we need look no further than the wisdom of David Bowie who described their stupefying effects as "a waste of time"[5], a reasonable substitute for the word "meaningless."

In Ecclesiastes 2:8 (NASB) Solomon describes two other principles that hold the nucleus accumbens in their thrall:

> Also, I collected for myself silver and gold and the treasure of kings and provinces. I provided for myself male and female singers and the pleasures of men—many concubines.

I discuss the role money plays in the addiction process in chapter 8. We will begin a consideration of the role of "love," its shallow casual quality manifested in Solomon's impulsive relationships with women, later in this chapter.

Serotonin Reuptake Inhibition

The media bring us regular news updates on the latest medical research. As depression is such an issue in western society, they often mention the various neurotransmitters involved in treating that condition. The two that have been targeted for decades are serotonin and noradrenaline (norepinephrine in North America.) More recently, focus has turned to glutamate signaling channels, with evidence that low doses of glutamate antagonists (like ketamine) have long-term effects in the relief of depression.[6]

For the moment we will focus on serotonin. As with many other compounds, serotonin has one function in the body and another in the brain. It was first identified in 1948 by Maurice M. Rapport, Arda Green and Irvine Page[7] due to its effects on muscle tone in veins and arteries. As with adrenaline receptors, serotonin receptors can cause blood vessels to contract, decreasing blood flow and raising blood pressure.

---

5. Hawking "The Collected Wisdom," para. 8.
6. Collins, "What You Need," para. 8.
7. Mack Whitaker-Azmitia, "The Discovery."

This function forms the premise for its name—being found in the blood serum (sero) and acting as a tonic (tonin).[8]

In the context of electrical signal transmission across the synapse, some drugs act as agonists as they directly activate receptors. Other drugs (inducers) cause the signaling molecule to be released by the upstream neuron. The effects of agonists and inducers are similar, but there is also a third approach that produces comparable results. This has to do with the efficient recycling of neurotransmitters. The upstream neuron contains only a limited supply of the signaling molecule (in this case serotonin); after an action potential has been propagated across the synapse, the serotonin molecules are reabsorbed into the upstream neuron. This is accomplished using a reuptake transporter, a protein complex that is similar in function to an ion channel, except that instead of allowing only a specific type of ion to pass across the cell membrane it provides this function for a single neurotransmitter molecule. Figure 3.1 illustrates how this is accomplished by a selective serotonin reuptake inhibitor (SSRI—for example Prozac) acting on the serotonin (5-HT) reuptake transporter.

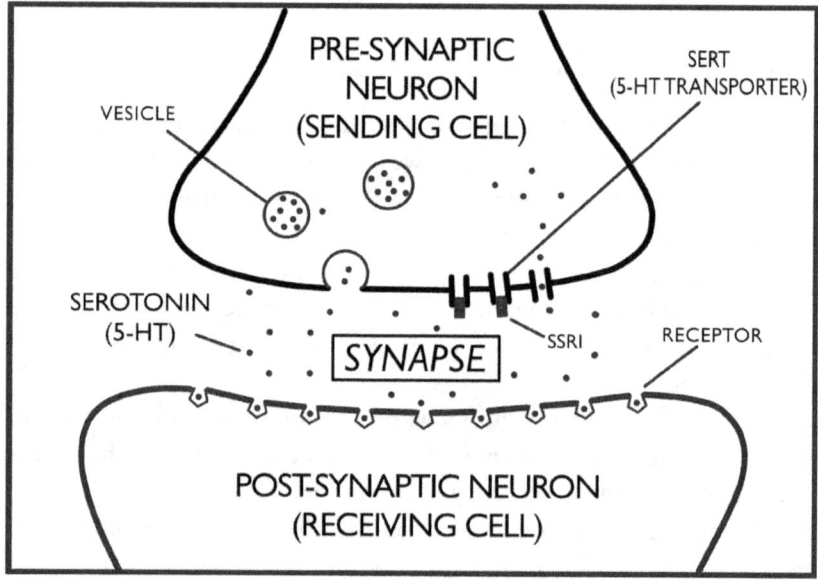

Figure 3.1 The Operation of SSRI Drugs. Wikimedia Commons.

---

8. Van Nueten et al., "Serotonin."

If this reuptake transporter is blocked, the neurotransmitter concentration in the synaptic cleft will remain high, and the downstream neuron will continue to fire. This achieves the same result as if the normal stimulus for firing the neuron had been received, but without requiring the stimulus itself. Most anti-depressants are serotonin-reuptake inhibitors and enhance serotonin signaling. Cocaine is a serotonin/dopamine/noradrenaline re-uptake inhibitor, although the case is still open regarding its additional function as a dopamine inducer. Experimental evidence indicates that the euphoric effects of cocaine cannot be accounted for by re-uptake inhibition alone.

Artificially Reproducing the Best Day of Your Life

While serotonin re-uptake inhibitors will ease the symptoms of depression for some, serotonin inducers (e.g., Ecstasy) might lift the individual to exhilarating highs. Serotonin inducers cause the vesicles storing the neurotransmitter in the pre-synaptic neuron to discharge into the synaptic cleft, resulting in unnaturally elevated levels of downstream electrical activity. This activity is the physical/chemical backdrop to the feelings of euphoric wellbeing as experienced by the spirit-observer. This level of activity might be appropriate in a normal setting on rare occasions like weddings, graduation days or when being given a clean bill of health following a cancer diagnosis. The key term here is "rare." In a normal setting only rare events will stimulate such a state, and only because they are perceived relative to the established baseline (or equilibrium) we could describe as the humdrum of everyday life. Thankfully, the baseline is adjustable, allowing us to bear (after a period of adaptation) such things as poverty, chronic sickness, and war. If the perception of such joyful events is really affected by neuro-chemical processes, it follows that we should be able to reproduce those phenomena by introducing the right drug into the system.

An illustration of a drug as the proxy for a rewarding life experience was presented in the May 2000 edition of Time Magazine under the title "The Lure of Ecstasy."[9] The article introduces Jack, an ecstasy aficionado who had used the drug over forty times. He describes the drug as making him feel like he did on the day he won a Rhodes Scholarship (a sensation produced by the release of endogenous chemicals.)

---

9. Cloud, "The Lure of Ecstasy."

Just think about that—ingesting a little pill can transport you back to the best day of your life within the span of about twenty minutes. Jack had been conveyed (by chemical means) to his experience of that special day forty times, which is not a good idea if only because it reduces the relative value of the good times.

Using a different example, imagine you purchased a winning lottery ticket. You would feel excited and euphoric about your stroke of good fortune. Those feelings of excitement are a function of electrical activity in the incentive and reward pathway, a state that can be reproduced chemically and predictably. You could easily reproduce the thrill of the win by ingesting the right compound, and this would be the chemical equivalent of buying a winning lottery ticket, except it would be a winning ticket every time! The elation and exhilaration experienced in both cases would be comparable. How much would you be willing to pay for a ticket that was guaranteed to win? It would certainly make sense to pay more for such a ticket. This is the kind of logic that drives the narcotic economy. It also indicates that there is a monetary value associated with a dopamine or serotonin induced reward. The price of a drug is dictated by the standard market forces of supply and demand, and demand is driven by the incentive side of the limbic loop.

While the examples above make intuitive sense, the vicissitudes of the mammalian brain throw us a curve ball. While a surface reading of the situation might assign more incentive to the purchase of a guaranteed winner, research has shown a much higher incentive-based dopamine response results when purchasing a ticket that *might* win. We will deal with this in more detail in chapter 9.

Depression

There has been much talk about the nature of depression. Is it a natural part of the human emotional landscape or a sign of dysfunction? We treat it as a sickness, a mental illness, and treat its symptoms, but doctors are less likely to consider what might be going on in your life than they are to prescribe a drug to treat the symptoms. A good doctor will encourage a depressed patient to seek talk therapy while they are taking the drug as this can help uncover the underlying reasons for their depressed state. There is evidence that the best treatment for depression is a combination of talk and drug therapy, but many people are happy to take an antidepressant and be done with it.

Which drug to take? Research shows that the average person with depression must try multiple drugs before they find one that works. This echoes the idea raised earlier that we all have slight differences in our genetic makeup that predispose us to a range of reactions to the same drug. The medical community is starting to address this problem by developing tests that detect specific genetic markers that indicate a particular drug treatment (personalized medicine.)

Antidepressants play a similar role in modulating mood to the ecstasy pills discussed earlier, except that instead of elevating mood to a once in a lifetime high in twenty minutes, they take a depressed mood and, over the course of a few weeks, return it to the normal range. How does depression arise? Imagine life is going well; we are happy in our relationships, our job and our family situation. We are content, a contentment mediated by endogenous chemicals that include serotonin, but suddenly things change. We are blindsided by a health crisis in our family; we lose our job, and the health situation introduces many stressors into the family dynamic. This results in changes in our brain chemistry which develop into a chronic mood disorder. We feel overwhelmed and miserable. We had no control over the onset of the health problem or the economic circumstances that caused our employer to lay us off, so those things might not be easily addressed. What we can change is our perspective on the situation, and that might be beneficial, but we could also take a drug that normalizes our neurotransmitter levels. It returns our brain function to the state it was in when everything was going well. The comfortable and positive life circumstances were a favorable environment that fulfilled a host of needs and we felt great. The drug is really mimicking our emotional reaction to the positive stimulus by creating an artificially satiated state.

The theory that depression stems from a chemical imbalance is controversial. We really do not know what causes depression, but we do know that selectively altering the chemical balance in the brain addresses the symptoms. That is why doctors continue to prescribe antidepressants, without knowing the root cause of depression. They are, for most people, an effective treatment.

The Search for Antidepressants with Selective Function

The drowsiness I experienced with Serzone, and amitriptyline is an undesirable side-effect of many anti-depressant drugs. In the 1980s the pharmaceutical industry sought a drug that would target serotonin selectively without the histamine effects. Research had shown that the antihistamine Benadryl (diphenhydramine) was a serotonin reuptake inhibitor and, therefore, a potential antidepressant, but Benadryl's blocking of the histamine receptor made it an impractical solution. The more selectively a drug binds with its target receptors, the fewer systems it impacts and the fewer side effects it exhibits. Taking Benadryl as a starting point in 1970, Bryan Molloy and Ray Fuller synthesized dozens of its derivatives, looking for one that would inhibit the reuptake of serotonin without affecting the histamine response.[10] The result was Prozac, which has almost no action on histamine receptors, making it a smart choice for people with antihistamine sensitivity. Despite the failure of Serzone and amitriptyline to address my mood issues, Prozac worked well.

The Use of Psychostimulants as Insecticides

All animals and plants have safety needs, and each species exhibits a set of self-protection mechanisms. Mammals possess the fight or flight response, which serves to remove them from dangerous circumstances or to neutralize the threat. This works well but would be ineffective if their legs, like the roots of a tree, extended twenty feet into the ground. Obviously, trees and other plants are incapable of movement and need to take a different approach to addressing threats. Some plants grow sharp thorns; others sting, while others ensnare their predators in sticky traps. A common strategy in the plant world is to produce, within their tissues, chemicals that affect their predators negatively when the plant is consumed. While prey animals are normally killed and eaten by other animals of superior size or number, the lives of plants are threatened by swarms of tiny insects. It is in their interest to dissuade their prey from eating them by making them feel sick. Chocolate is one plant which shares the ability to produce caffeine (as an insecticide) with five other plants.[11] Tobacco produces nicotine for its insecticidal properties, a feature that was used by generations of humans who used it to kill six-legged pests in their farms and gardens. While these two compounds are toxic to insects, they are not (in small quantities) harmful to humans; they do, in fact, act as psychostimulants.

10. Wikiwand, "Fluoxetine," para. 43.
11. Gunter, "6 Plants."

What would the working world do without its fix of coffee in the morning and at regular intervals throughout the day?

It is not that the insect brain is drastically different at the chemical level to that of the human, but while these stimulants have positive effects on us, they have rather deleterious consequences for creatures of the six-footed kind. What better way to dissuade your predators from eating you than by rewarding them each time with a very unpleasant experience (a bad trip)? Certain acacia trees native to the state of Texas take this to a whole new level. Their tissues contain amphetamine and methamphetamine as predator deterrents.[12] This is especially surprising because these compounds were discovered in the lab long before they were thought to exist in nature. In the same way, the Coca plant produces cocaine as an insecticidal neurotoxin.

The Chemistry of Love

In addition to containing caffeine and the psychostimulants theophylline and theobromine, as predator deterrents, chocolate also contains phenethylamine (PEA,) also known as "the love drug."[13] PEA is produced naturally in the human brain and reaches elevated levels during that most sought-after state called *being in love* (Greek *Eros*.) It acts as a neuromodulator in the brain (a Trace Amine Associated Receptor 1 [TAAR1] agonist) and like serotonin it is associated with controlling muscle tone in the blood vessels. Science has identified this compound as the main ingredient in a wide array of substances occurring in the body and the lab. These drugs each add extra components to the PEA molecule and are referred to as substituted phenethylamines. This knowledge has practical application to our human experience because the neurotransmitters adrenaline, noradrenaline, and dopamine are all substituted phenethylamines. As such they possess comparable properties to those of PEA, the love drug. Dopamine inherits the blood vessel constricting properties of PEA. In its pop-culture context, many episodes of the medical drama series ER showcase the doctors' use of dopamine to treat low blood pressure.

As we have already noted, the BBB prevents messaging chemicals (like PEA and dopamine) from entering the brain and interfering with

---

12. Wikidoc Contributors, "Acacia berlandieri."
13. Schwarcz, "The Chemical of Love."

normal function, but this is not the case with many PEA related drugs. A major issue in our scientifically enlightened society is the abuse of such drugs specifically because of their molecular structure and how this affects signaling within the brain. As we will see, this is always (either directly or indirectly) related to dopamine function.

Amphetamine

The compound with the highest pop-culture profile, in terms of driving a euphoric high, is alpha-methyl-phenethylamine. Let us break that name down into the acronym by which it is best known:

*a*lpha-*m*ethyl-*ph*en*et*hyl*amine* = Amphetamine

Amphetamine is commonly available as a treatment for narcolepsy (unwanted bouts of falling asleep,) attention deficit disorder (ADD,) and obesity. It is a dopamine inducer—it causes the upstream neuron to release dopamine into the synaptic cleft. If an extra methyl group is attached to amphetamine, it becomes methamphetamine, which has the same cognitive effects but is more easily absorbed by the fatty tissues of the brain. Some jurisdictions (such as the United States) allow the prescription of methamphetamine (Desoxyn) for the treatment of the same three conditions mentioned above.[14] Desoxyn has the same formula as crystal meth, albeit in a less robust form that cannot be as easily absorbed by the body. Many doctors are aware of the abuse potential of these drugs but prescribe them, believing that the positive effects will outweigh their habit-forming tendencies. As with every drug, some people will like amphetamine and be able to handle it; some will not like it, and others will become engrossed with its euphoric effects.

Ritalin

I can speak to this principle from my own accidental experience with another PEA derived substance—Ritalin. I was diagnosed with Adult ADD at a time when Ritalin was tightly controlled in my local jurisdiction. To access it you needed a triplicate prescription, one with a top sheet, and two sheets of carbon paper. The top sheet is given to the patient, and the other two are used for tracking the patient's consumption

---

14. WebMD, "Methamphetamine."

of the drug across all pharmacies overseen by the college of physicians. In 2004 I was informed that the triplicate requirement for Ritalin was being removed, but consumption would still be tracked in a central database. The doctor added "I don't know why it was ever classified as a controlled substance in the first place. . . it's harmless." This is not the opinion of all doctors, however.

A few months later I attended a job fair in another much larger city but accidentally left my Ritalin at home. As I wanted to be on-point for interviews I went to a local walk-in clinic to get a new prescription. As soon as I mentioned the name of the drug, the doctor excused himself and left the room. Returning a few minutes later, he invited me into a different room where a second doctor was busy looking at my Ritalin prescription history at a computer terminal. They asked me to confirm some details from the record to determine if my request for a prescription was authentic. After they were satisfied everything was in order, the first doctor gave me the following admonition:

"Did you know there are people within two blocks of here that will kill you to get their hands on that drug? They inject it when their supply of heroin is interrupted. You'd better keep it in a secure place at home to avoid the scenario where you have a guest at your house who checks out the cabinet in the bathroom and finds it. They tell the wrong people about it and before you know it your house has been broken into and the Ritalin is gone."

Hearing this was at bit of a shock. Apparently, the reason for the first doctor's immediate exit was his concern I was a threat—he wanted to make sure I was outnumbered!

Is There a Clear Distinction Between Legal and Illegal Drugs?

This story introduces the controversial distinction between legal and illegal drugs. It is in the best interest of law enforcement agencies to promote this distinction, but it is often artificial. If a heroin addict can also use the legal drug Ritalin (albeit for an illegal purpose,) then what is the difference? It comes down to whether the drug has been deemed, by government regulators, an appropriate treatment for a recognized medical condition. The distinction becomes less clear with the opioid pain killer fentanyl. In today's illegal drug market, a high proportion of what is being sold as heroin on the street is fentanyl, a legal opioid pain killer that is used in hospitals and for outpatient treatment. If the effects of fentanyl were materially different from those of heroin, I doubt this substitution would pay off for those who deal. The fact is that heroin and fentanyl both activate the mu opioid receptor in an analogous way and with similar effects, although fentanyl is approximately twenty times as potent as heroin. The mu receptor stimulates the release of dopamine in the nucleus accumbens by its action on interneurons in the ventral tegmental area. Ritalin does the same through its action as a dopamine reuptake inhibitor.

Ritalin has its own evil cousin, cocaine. Although cocaine is not derived from PEA, they both have similar effects on dopamine signaling. In that role Ritalin is even more powerful than cocaine.[15] The similarity in subjective experience between them has also been studied. In one case a group of habitual intravenous cocaine users were enrolled in a double-blind study to determine their ability to distinguish between the two. In about half the cases, the subject misidentified cocaine as Ritalin or Ritalin as cocaine, even though they had years of daily exposure to the latter.

Ecstasy

If we add another set of compounds to methamphetamine, we can create 3,4–methylenedioxy–methamphetamine, better known as Ecstasy or Molly. This is the popular party drug also known as the hug-drug. It increases feelings of love and empathy by inducing the release of serotonin from the upstream neuron. This is one feature of PEA derivatives that is seen repeatedly—their ability to create states similar to "being in love." It is hardly surprising that a drug that recreates the natural experience of emotional connection and feelings of worthiness would be so engaging. What is more, it is delivered in a much more predictable way and often on a more reliable timeline than a real relationship and without the drama and baggage that might be involved in the latter. Needless to say, there are some serious side effects to abusing PEA based drugs, like cardiac arrest, stroke and death.

PEA Based Pharmaceuticals

PEA based drugs are also used to treat other conditions. Here are some drugs you may have heard of without realizing how they relate to drugs

15. West, "Chidren's drug."

of abuse. Prior to 2015, if you used a Vic's Inhaler as a nasal decongestant, you would have been inhaling methamphetamine.[16] Although it is only the levo stereoisomer (the left-hand isomer) of methamphetamine (which is less psychoactive,) it releases norepinephrine on contact and activates adrenaline receptors in the lining of the nose. This quickly reduces inflammation. Levo methamphetamine is also psychoactive enough that people have been known to swallow the cotton pad inside the inhaler to get high.

If you are trying to lose weight, your doctor might prescribe phentermine as a PEA based stimulant. Bupropion (Wellbutrin or Zyban) is a keto-amphetamine that is used as an anti-depressant and smoking cessation aid. Two side effects of Wellbutrin are worthy of note: it increases the intensity and vividness of dreams and lowers the seizure threshold, causing the occurrence of absence seizures in some people with no previous history of epilepsy. Some of these drugs target the brain and have body side effects; for example Ritalin will quickly clear a blocked nose. Others target the body and have brain side effects, like the Vic's Inhaler.

King Solomon's Folly

With the allure of PEA derived substances as our backdrop, it is clear Solomon had a problem with that family of drugs. This was expressed through thoughts and behaviors rather than by their administration from external sources. The drugs that ensnared him were produced within his body and were allowed to proliferate through his lack of moderation with women and money. After all, having 700 wives and 300 concubines, and bringing 666 talents (twenty two tons) of gold into the treasury each year is a little excessive. These are the things that occupied his thoughts, rather than the values he had pledged to live by. Moses was aware of this snare for future kings when he wrote in Deuteronomy 17:17 (NIV):

> He must not take many wives, or his heart will be led astray. He must not accumulate large amounts of silver and gold.

While God ultimately rejected Solomon because he worshiped the gods of his foreign wives, it was the wives' allure that led him to that point. This is a lesson for the addict—the drug is rarely the worst problem; it is the

---

16. Smith et al., "Methamphetamine."

second order effects (destruction of relationships, job losses, bankruptcy, homelessness, and imprisonment) that have the most impact.

In summary, two central themes of this book are the role of dopamine and other substances in establishing inter-personal connection and the place of drug abuse as a strategy to mimic connection. As the work of shame and vulnerability researcher Brené Brown has shown, "We are psychologically, emotionally, cognitively, and spiritually hardwired for connection, love and belonging."[17] Viewed through this lens, drug abuse can be seen as a love-disorder, and love can involve endogenous drugs of abuse.

Thematic Takeaways

Feelings: The feelings we experience are often driven by the release of endogenous hormones and neurochemicals. These are emitted in response to higher level cognitive functions, perhaps fear, the appreciation of beauty or our perception of our own value. The feelings themselves are perceived by the spirit observer (our own conscious awareness) which watches over our every moment. Drugs that enhance serotonin function promote feelings of happiness while those promoting dopamine function enhance perceptions of pleasure.

Neural Activities: Depression is commonly treated using SSRI drugs. These raise the level of serotonin in the synaptic cleft causing the downstream neuron to fire persistently. This can potentially lift the fog of melancholy and shine a metaphorical light on the gloom of depression over a period of weeks. Conversely, serotonin inducers (like Ecstasy) cause the upstream neuron to release serotonin into the synaptic cleft, triggering an immediate and more intense experience but with similar qualities. Amphetamines are dopamine inducers that bind with the TAAR1 receptor, causing dopamine to flood into the synaptic cleft and promote enhanced feelings of wellbeing.

Relationship: PEA stimulates many of the positive feelings associated with love, feelings that play a central role in establishing human relationships. PEA turbocharges dopamine function in the short bursts that punctuate the obsessive state of infatuation at the start of a romantic relationship. As both PEA and amphetamine act by binding with the TAAR1 receptor,

17. Brown, *Daring Greatly*, 68

the rewarding experiences triggered by both have many similarities. You can think of PEA as the spark, dopamine as the fire, and oxytocin as the glue that holds relationships together. The amphetamines, and other drugs that promote dopamine function, are unwittingly used by many as stand-ins for the rewarding relationships they lack.

What you can do now: Consider whether the substances or behaviors you are using are playing the role of relationship in your life.

4

Thinking Outside the Box

"To me thinking outside the box means: crossing disciplines and pulling in expertise and perspective from outside of the standard boundaries." —Dror Benshetrit

THE NATURE OF CONSCIOUSNESS aside, we must agree that relational connectedness is a fundamental component of everyday life. But no matter how connected we feel in our relationships, we still operate as islands of awareness. We may see things from another's point of view, but we can never *be* them. We can never feel the actual sensations they experience in their interactions with us. Moreover, their perspective on the world may diverge radically from our own. The way they see the world (their *worldview*) is a synthesis of the knowledge, experience, and wisdom accumulated over a lifetime with some genetic influences thrown into the mix.[1] The nature of worldviews themselves is open to interpretation. There are those that classify them in terms of neural development, some who see them as a function of cultural evolution and others who see them as rational perspectives.

Trying to navigate a system of this complexity without a map presents challenges. Having driven across London (UK) many times in my youth and getting lost each time, I am familiar with this rudderless state of being. The trip always began in confidence, but at some point in the journey doubt transformed into perplexity. In those situations, the map saved me from wandering endlessly from intersection to intersection.

1. DeAngelis, "Are beliefs inherited?"

In a search for clarity, predictability, and understanding, humanity has always sought to map their emotional, spiritual, cultural, social, and physical realities. Without a map it is impossible for us to know where we stand in relation to the competing influences of life. Although my adolescent cross-London treks were challenging, at least I had a destination in mind—a luxury that often evades us in the wider context of life's journey. Maps are especially useful in providing a bird's-eye view of the territory we traverse, allowing us to perceive structure and form in a chaotic landscape. Religion is a guide to the topography of our spiritual space, while other domains of consciousness fall under the psychological, social, and physical sciences.

To date these spaces have been classified into the following categories:

Perspectives on reality

Stages of personal/cultural development

Lines of intelligence, capability, and achievement

States of being

Personality *types*

While these exist as independent vertical fields of study, they have also been classified horizontally, by philosopher Ken Wilber and others, using an *Integral* approach.[2] As its name suggests, Integral Theory seeks to *integrate* the attributes of each field by identifying where they intersect. The relationship between spirituality, addiction, and integral theory will become clearer as we proceed, but I want to start by exploring the concept of the *worldview* and how it relates to our self-concept.

The Zeitgeist

The world view embraced by the majority in a culture forms the zeitgeist, the spirit of the age, and as such it is that by which "normal" is defined. A person can veer away from this view in one direction or another only so far before the gravity of the zeitgeist attempts to pull them back in line. People enjoy the unity of conformity and the societal acceptance it affords. They want to be liked, they love peace and acceptance, and those are some of the ways the zeitgeist exerts its influence.

2. Duffy, "A Primer."

When the zeitgeist pulls, we may *allow* our forward direction to be determined by external forces, like status and prestige, wanting to comply with societal expectations. The gravity of the zeitgeist is powerful, and as scientists use gravity to direct a path from earth to distant planets, so we can exploit contemporary cultural forces to our benefit. Some allow themselves to be drawn towards the comfort-zone of the cultural center, while others use the culture's gravity as a slingshot, propelling them in new and life-serving directions.

Those on the fringes of normalcy, the outliers, may move in stealth while hiding their true motivations or they may be authentic participants in their own stories. They may be con-men or sweethearts, but because of their divergence from the norm they are all viewed with suspicion. For those on the fringe, the cultural center is unpredictable and repellant. The in-crowd sees adherents to other worldviews as narcissistic and deceitful. Truth be known, the outliers merely possess a different perspective on reality from that of the majority.

From the integral perspective, philosophy, psychology, and religion are tools that help us navigate this space. These are lenses that refract the light into discernible images, ones we recognize within ourselves and within community. An inability to see these images is a form of blindness, as it is by them that we navigate our social realities. With these images as our maps/guides we can embrace our value system and enter a place where a commitment to the things we hold most dear supersedes our longing for acceptance, approval, and recognition. This is the realm of self-actualization, the process of "becoming real"[3] of setting aside all secrets and masks for the sake of integrity. We will never find the door to this place without the clarity provided by our helpers, be they mentors, family, or spiritual guides. Before stepping into integrity, we are unsure of our purpose and struggle to find meaning, but once within, the scales fall from our eyes. Those who take this step before their twilight years are blessed. Without it we tend towards self-destructive coping strategies to help us survive.

---

3. Kaufman, "What Does It Mean?"

The First, Second, and Third Persons

Emmanuel Kant, the father of modern philosophy, categorized human perspectives into three classes.[4] These correspond with three facets in the lens through which we see the world. Because we first use them without training the images presented are blurry and indistinct, but with help we can bring them into focus. The first perspective is "*I.*" In our role as observer we are operating as an individual. The knowledge of our inner states of being, thought, and sensation are subjective. If we were to describe them, we would preface our comments with "I." I feel tired, I am in pain, I know such and such.

Culture and community view the world in a way that serves them. The topics of discussion within a community are inter-subjective and coalesce around the central idea of "*We.*" We have these beliefs, we have these goals, we have these values.

Separated from both by the great divide between subjective and objective experience is the realm of "*It.*" From this perspective all things are measured by their physical attributes. Size, brightness, temperature, location, and function. It has a length of twelve centimeters and a width of fourteen centimeters, it is green, it weighs ten kilograms. These are all particulars of the it perspective, but this perspective also has a relationship with the other two perspectives it *observes*.

The *it* perspective never communicates with the other two because it does not have to; their attributes can be ascertained by external observation alone. Systems can be imposed on the intersubjective so that cultures and communities are kept healthy and safe. Individuals can be kept healthy by performing assessments with medical imaging or physical examination. The *it* perspective does not ask or empathize, it just tells. It sees and speaks in monological terms and the monolog by which it directs is authoritarian. It dispassionately measures location, form, and functional fit. The one being observed provides no input to the decision-making process and it cannot touch the humanity within the observer, as there is none. Wilber refers to this assessment process as the monological gaze.[5] The *it* perspective can be applied to the subjective and the intersubjective. From the it-space we can say of the individual—it has a blood pressure of 120 over eighty, it has a fracture of the tibia, and of the

---

4. Wilber, *A Brief History*, 243.
5. Wilber, *A Brief History*, 110

intersubjective—it requires a health care system, or it needs a judicial system for its ordered existence.

Limitations of the *It* Perspective in Healthcare

The Stethoscope provides a powerful example of this kind of objective assessment. It was the first medical tool to be employed in assessing a subject's physical state without requesting information from them directly. The physician could now hear the tell-tale signs of a heart murmur or an arrhythmia. They could diagnose bronchitis or pneumonia without asking the patient "how do you feel"? or "does it hurt when I do this"? This was a turning point in the evolution of medical practice. The addition of a new diagnostic tool was an advancement for the medical profession, but along with the diagnostic capabilities of the tool came the tendency to objectify the patient. With the option of making an objective assessment of a patient's state the doctor might interpret the absence of observable symptoms as an indicator of healthy function.

The objectification of the patient is costly in terms of missed opportunities for healing. Patients are sometimes released untreated from hospital emergency rooms because the cause of their symptoms cannot be identified. It is almost certain there is a physical problem that could be alleviated by the correct diagnosis, but the doctor's conclusion is sometimes "it's all in your head." The *it* perspective can lead to over-confidence on the part of the medical practitioner. Conversely, it can be used with compassion and promote connection, but only if used empathically by including the *we* perspective.[6]

The AQAL Matrix.

Wilber takes Kant's concept of the big three (the subjective, inter-subjective and objective) and adds a fourth perspective, presenting the concept graphically as The All Quadrants All Lines (AQAL) matrix. The matrix is divided horizontally into the individual (top) and collective (bottom) and vertically into the subjective (left) and objective (right.) This simple visualization portrays "four faces of spirit," the four ways in which consciousness can inhabit reality. It is a spiritual map. We can overlay this matrix on the human mind and bring out its features with amazing depth and clarity.

---

6. Germa, "Stethoscopes and Stories."

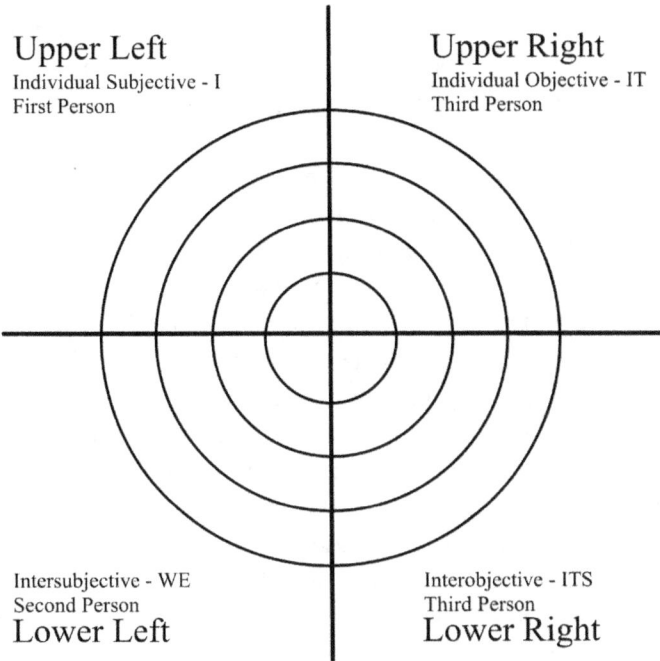

Figure 4.1 The AQAL Matrix.

When our world view is directed by one of these four perspectives alone, it is easy to view external reality through only one facet of the lens. This tendency is summed up in the idiom "when you have a hammer everything looks like a nail." To avoid this, we will dip in to all four quadrants in our consideration of addiction, but my focus will be on integrating the upper left and upper right. The upper left is the individual subjective. It encompasses all features of our internal experience, thought, emotion, pain, suffering, loneliness, pride, ego, and belief. The upper right is the objective view of the individual and describes all attributes of human physicality. This is the perspective from which the stethoscope, X-rays, MRI, CT and PET scans operate. It is also the space where all physical features, functions, and dysfunctions are described. These may be chemical or structural in nature and tests may be performed on bodily tissues, fluids, and gasses to assess the individual's internal state.

The upper right quadrant has no heart. It cares nothing for feelings or needs and is only concerned with measurement and classification. It classifies the physical structures of the body, the organs, and their operation, the chemical cycles of respiration and the nutrients, and transport methods that support them. It can describe the electrical currents underlying emotional and transcendental states, but it says nothing of the subjective qualities of those states.

We have already considered some limitations of the *it* perspective, but it also contributes many positives. Diagnostic tools far surpass the individual's ability to detect internal states through introspection alone. Medical assessments are a vital tool in diagnosing a patient's condition. They highlight the traditional style of external observation, a unidirectional assessment of a subject by an external observer. There is also an alternative—to look from within the subjective realm towards the objective and measurable. This involves the mindful exploration of our internal states and resolving the question "what is the bio-chemical/bio-physical cause for this state"? You might wonder how this could be of use? I first used this approach due to my burning curiosity about my own emotional states and how these might be expressions of biological activity. It is my contention that this question gives us leverage in our struggles with addiction by transforming the feelings that drive us into observable entities.

Defining Addiction

The medical community defines *addiction* as the habitual, self-harming use of activities or substances for their limbic effects.[7] Seeking limbic stimulation on its own is not the problem, as our existence as a species depends on it. Our issues with addiction stem from the question "*Why am I seeking pleasure and fleeing pain*"? We are all experts at seeking out what makes life more enjoyable and resisting the averse. We engage in activities because we enjoy them and afterwards look back on them as rewarding. A weekend of skiing in the mountains, last night's win by our favorite team, the concert we attended. Each of these experiences is the precondition for a discrete and rewarding state of being, but such states may also occur for no apparent reason. For the curious sort, the distinction between predictable and unpredictable emotional states allows us to interrogate the objective. We can ask it "Why do I feel this way"? We lock eyes with the monological gaze when we seek a neuro-chemical

---

7. Rodríguez and Navarro, "Role of the."

explanation for a felt-sense experience. So, what can the *it* perspective contribute to our understanding of subjective states?

A Winter's Glow

My first foray into reconciling my own felt-sense experience with objective scientific knowledge could be termed "a winter's glow." I'll use this as an example of how you might consider your own internal states in the light of easily accessible science. In my younger years I had noticed, over successive winters, a sense of well-being that occurred somewhere between the onset of fall and springtime. I realized later that these states were always accompanied by a case of the common cold. I considered various explanations for this, for example, because I always had a cold at Christmas time when I was a child, had I associated the condition with feelings of Yuletide euphoria? My doctor suggested that sickness allowed me to lower the expectations I had of myself around work and the other responsibilities of life, allowing me to relax. These explanations failed to turn up a satisfactory answer to my conundrum and my curiosity remained.

Then I realized that every time I have a cold, I take cold remedies. Was a constituent of the cold remedy responsible for this emotional state? It did not take long to identify the cough suppressant dextromethorphan (DM) as the culprit. DM is primarily a general anesthetic, but in small doses it works well as an antitussive (it interrupts the cough reflex.) I wondered what other function it might have that would explain my experience more broadly. A little research revealed that DM is the dextro (D—*right hand isomer*) form of the synthetic opioid levomethorphan (L—*levo* meaning "left hand.") That caught my attention, mostly because of the place that other opioids such as heroin and fentanyl play in our shared cultural basement. My sources assured me that DM did not act like an opioid because the molecular structure was a mirror image of LM. It was, however, a hallucinogen, a stimulant like phencyclidine, a depressant like laughing gas and a serotonin re-uptake inhibitor like Prozac. Its antitussive effect stems from its action on the two sigma receptors. Through my later experience with Prozac, I concluded that the DM-induced glow was a result of serotonin reuptake inhibition or sigma receptor agonism.[8] This similarity between Prozac and DM contra-indicates

---

8. Nguyen et al., "Involvement of."

their co-administration as this may result in serotonin syndrome, a condition that results from extreme nervous-system stimulation.

A Creative Mind

You might find this discussion overly materialistic, especially in the context of a book on spirituality, but my goal is to reconcile the spiritual and material aspects of consciousness. The most productive way I know of accomplishing this is by using observation and the scientific method. In a world of rationalism, our culture teaches us to value science above all. We read about the immutability of scientific fact and often in the same breath the impossibility of the existence of a God. But this shows a false perspective regarding science and its relationship to the mind. The upper right quadrant of the AQAL matrix and its monological gaze sees only those things that are detectable with today's tools. It exists in the physical space. It attempts to ascribe features, attributes, measures, and location to the physical components of the body. But what about the mind? The mind is constantly in motion. It is not a physical feature. We can measure the physical attributes of its operation such as the Alpha, Delta, and Theta brain waves detectable using an EEG. We can produce visualizations of the flow of blood to distinct parts of the brain using Functional Magnetic Resonance Imaging and can extrapolate from our experimental knowledge of brain morphology which functions might be active based on the results. We can attach radioactive tracers to targeted drugs and see where those drugs are most active to determine the messaging profile of an individual brain. From that we can attempt to explain some feature of behavior or subjective experience, but we can never put our finger on what the mind is. I decided to be a phenomenological observer of my own internal states to widen my understanding of the mind's operation, something you can do also. Thankfully the research I base my findings on has already been done for me.

The Scientific Method

The scientific method involves several steps:

1. Generate a proposition or idea (a hypothesis)
2. Conduct observations of physical reality to test the hypothesis

3. Consolidate these experimental results to construct a theory
4. Continue experimentation to either prove or falsify the theory over time

Using this approach Charles Darwin reasoned that the adaptations that gave a new form of life advantages over the old would be progressively passed on to later generations.[9] He called this evolution by natural selection or survival of the fittest. He also proposed that the fossil record would reveal, through further research, the missing forms at each point in the adaptation of a species. When this proof had eluded scientists for several decades, the evolutionary biologist Stephen J. Gould proposed a further enhancement of the theory that he termed "the survival of the luckiest."[10] It is true that a creature which experiences an advantageous mutation could be killed in a rock fall and be unable to pass on that mutation to its progeny, but this is certainly not something that can be supported by scientific evidence. It is a thought experiment.

It is not that the thought experiment is without value, it is just that it is a component of hypothesis-building, and not the direct foundation of a theory, something that can only be supported by direct observation. To illustrate the validity of the thought experiment in scientific research let us consider the methods employed by one of history's greatest minds, Albert Einstein.

Albert Einstein—An Out-Of-The-Box Thinker

Einstein had a particular predisposition towards science—he found theoretical analysis much more rewarding than experimentation. In university he was censured for his poor attendance record in chemistry labs. He was also known to author research papers with only theoretical and mathematical underpinnings, and to conclude them by eliciting help from others in testing his hypothesis rather than publishing the results of his own experiments.[11] Many of Einstein's thought experiments and theoretical propositions are notable for standing the test of time. Most recently, as of this writing, the existence of gravitation waves has been confirmed.

---

9. Gregory, "Understanding."
10. Gleick, "Survival of the Luckiest."
11. Isaacson, *Einstein*, "The Light-Beam Rider," para. 23.

One thing Einstein refined while developing his general theory of relativity was his use of mathematics, not just in describing the behaviors of physical phenomena, but also in exploring attributes of reality that had not been previously observed. He was able to create such an accurate representation of reality in his mind, by thought experiment alone, that those ideas are employed today in every GPS device on the planet including your smart phone. The thought experiment is something we all employ in our interactions with the world. It is what allows us to accomplish feats we have never attempted before. It is also a tool I used extensively when compiling the content for this book.

The thought experiment frames the universe as an intelligible system (one that can be understood) and indicates that intelligence itself can accurately predict aspects of reality that are unknown, unobserved, and even unobservable. Einstein was smart, but his performance in university was not stellar.[12] He did, however, have two attributes that allowed him to achieve supremacy in his field: an open mind and an obsessive curiosity about the physical universe. Unlike his predecessors in the world of Physics, those who had firm beliefs based on prior research, he was young and free from attachment to a particular cultural understanding of the physical world. I say cultural, because the culture of a scientific community has a powerful influence over what can be researched, hypothesized or even thought. If a proposal is made to the community that contradicts the accepted view it will be resisted. Scientists tend to self-censor their ideas to fit within accepted norms, as we all do in matters of reason. This is the gravity of the zeitgeist of that culture at work. Another feature of Einstein's personality that led to his success was his strong non-conformist streak. He was not beholden to the scientific zeitgeist and thought well outside the box. This, in combination with his obsessive interest in the physical world, are not features we associate with intelligence, yet they were vital strengths that allowed him to excel in his scientific endeavors.

Cultures And Sub-Cultures

The zeitgeist exists in the lower left quadrant of the AQAL matrix, the inter-subjective. This is where subjects interact and build cultures. These cultures constitute systems of behavior, morals, contextual knowledge, principles of aesthetic perception and belief systems. There are strong

---

12. Isaacson, *Einstein*, ch. 3, "Graduation, August 1900."

interactions between the subjective and intersubjective quadrants. The whole of subjective experience is interpreted, by the individual, in terms of the culture. Culture defines the parameters of value judgment, self-esteem, law, and order, the semantics of language, and many other conventions. The dictates of the culture are so powerful that meaning itself is culturally defined.[13]

Each culture contains sub-cultures. Workplaces, churches, hobby clubs and sporting organizations each have their own cultural norms. These cultures usually draw on the culture at large, but there may be some tension between them, as is evidenced in the relationship between law enforcement agencies and motorcycle clubs. While motorcycle cultures usually define a strict value system and code of conduct, those values differ from those at the cultural center (traditional/modern), prompting continual suspicion between the two. Motorcycle clubs operate using the more basic elements of the human psyche (primarily *the will*) and law enforcement agencies pressure them to conform to a set of rules designed to cope with a more complex set of living conditions, where the needs of all are considered. If we move up one level in the zeitgeist spectrum, from traditional to modern, we find tension between law enforcement, in the form of government regulators, and capital markets. Regulators enact policies that protect investors but limit the potential gains for the securities industry. Governments are subjected to intense lobbying by financial services organizations which leads to cycles of deregulation and financial crisis, followed by bailouts and regulatory retrenchment. Spanning these competing spheres of interest is the realm of political discourse. In simplistic terms, these three perspectives can be described as Heroic/Warrior, Traditional, and Modern, as shown in figure 4.2. In this figure, cultural evolution progresses from left to right.

---

13. Kennedy, *Beyond Race*, 1–12.

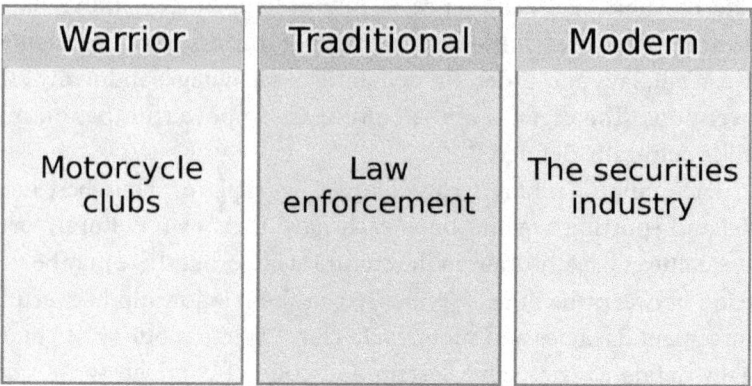

Figure 4.2 Three Twenty First Century World Views.

Because meaning is judged in cultural terms, inter-cultural relations at the national level can also be problematic. An authoritarian culture may ascribe meaning to discipline, both self-discipline and the discipline of corporate control. Other cultures which value empathy, rights and freedom may be judged by such a culture as weak and inferior while they see the authoritarian culture as dominating and inhumane. Either way there is a clash in the metrics of meaning which affects the relationship negatively. In our recovery this friction will arise, not only with our associates (those who share the recovery journey with us,) but also between various aspects within ourselves. The integration of these "subjects" within us is an important part of the recovery process.

The Genius of Childhood

One of the challenges we encounter in our internal struggles with addiction is the tendency towards mental inflexibility over time. This helps explain why scientists perform their most valuable research in their early years. Einstein himself once said "a person who has not made his great contribution to science before the age of thirty will never do so." For him, this turned out to be true. His perspective evolved from that of a free thinker to a more conservative one over the period from 1915-1930, by which time the field of quantum-mechanics had matured. He did not appreciate the fuzzy nature of quantum theory and its reliance on probabilities to explain the operation of matter at the atomic and subatomic level. He believed the kind of consistent reasoning he employed in his theory of relativity was a universal principle operating at the quantum level, the level of human experience, and on the vast scales of interstellar space. It was in this regard that he made his famous statement "God does not play dice." Einstein called the task of reconciling the tiny with the gargantuan *the unified field theory problem* and spent the remainder of his life trying to address it, without success.[14]

Were he with us today perhaps he would still be hammering away at that task, meanwhile mainstream physics has kept on down the rabbit hole of quantum theory, and so far, it appears that God does indeed play dice. It was that one belief more than any other that stymied Einstein's progress. His intellectual antecedents were blinded by other beliefs and assumptions, ideas that the young Einstein was not constrained by, but the ineffectiveness of their ideas to fully describe the observable universe left the door open for him to take the next great conceptual leap forward.

The Constraining Influence of Knowledge

This shift from open mindedness to a restrictive belief-based system that muffles perception and stifles creativity is something that we all experience as we mature from children to adults. Parents see it in their children[15] when they (the parent) are struggling with a video player, computer, or smart phone. The child becomes impatient and says, "let me do it," then within a few seconds the problem is solved! It is as if our cognitive abilities are locked in place unable to exhibit elevated levels of creative flexibility. However, within our limited sphere of expertise we perform on a vastly superior level to that of our children. Our thoughts travel along paths rutted so deeply that the wheels of reason are unable to buck their way free to follow other more effective routes. We can be confident, however, that the path we have habitually followed will take us to our destination. This is the great strength, and weakness, of education.

This shows that a community outsider may be as qualified as an expert within a particular field to make judgments concerning that intellectual space, because their minds are not constrained by the same forces. Anyone who has the inclination to educate themselves in a field

---

14. Isaacson, *Einstein*, ch. 14, "Newton's Bucket and the Ether Reincarnated."
15. Quora, "Why are younger people."

of choice may well have a valuable contribution to make. This line of thought is supported by a research paper[16] that explores the paradoxical ability of novices in a field to outperform intermediate practitioners in decision making. If Einstein, one of the smartest men who ever lived, became entrenched in his thinking, there is no reason to judge ourselves for doing the same regarding our addictions. Neither should we judge ourselves as incapable, as self-taught novices, of solving hard problems given the right information. Do not think you need to be a trained psychologist to help yourself or that you need to be a neuroscientist to draw on that field of medical expertise (or even author a book!) You are free to indulge in the field of Citizen Science.

Denying Our Addictions

I encourage you to think outside the box concerning your own self-destructive behaviors and not to be held back by habitual patterns of thinking that might include *limiting beliefs*. Denial is also a constant companion of addiction[17] which draws strength from the ruts our mind inhabits and from the fear of losing the strategies we have invented for our own emotional protection. As our lives progress and we are subjected to various trials, we invent ways to avoid the pain that flows from these experiences. In doing so we *educate* ourselves regarding the nature of life and how to survive it. We typically latch on to the first strategy we find, at an early age, but neglect to reassess this later in life. We internalize these strategies and before long they become part of our identity despite the damages they cause.

Denial is a form of blindness and the mind beats social media as the most effective echo chamber ever constructed (it is a social media platform with one member!) We listen to our own thoughts continually and most of the time we believe them, even when they make no sense. They form the ruts in which the wheels of reason remain trapped. Without input from the things we read, listen to, watch, and from those who care about us, we will not see new perspectives and will remain trapped within our minds. Become an observer and critic of your own thought processes and admit that they might not be rational. But as you explore your own internal thoughts and states, through the research principle

---

16. Herbig and Glöckner, "Experts."
17. Heshmat, "The role of Denial."

of compassionate curiosity, be kind to yourself, forgive yourself for past wrongs and let go of the self-blame. Do not be an old scientist who is set in his ways but be flexible, accepting that there will be a cost to change but things of great value are waiting for you.

Changing Our Way of Thinking

As Jesus said in Luke 5:37–39 (NASB)

> And no one puts new wine into old wineskins; otherwise the new wine will burst the skins and it will be spilled out, and the skins will be ruined. But new wine must be put into fresh wineskins. And no one, after drinking old *wine* wishes for new; for he says, "The old is good *enough*."

Old wine is lifeless, and old wineskins are hard and inflexible like the implacable mind. New wine is alive, pushes against its container and stretches it into a new form. The new wine (way of thinking) is unpalatable to those who are attached to the old and breaks their view of the world and themselves. You may wonder how it is possible to store new wine in old wineskins without destroying them? This is the great challenge of recovery and there will be some damage to the pre-existing structures you have created for your own protection. But hope lies in the words of Paul in Romans 12:2 (NASB):

> And do not be conformed to this world [zeitgeist], but be transformed by the renewing of your mind, so that you may prove what the will of God is, that which is good and acceptable and perfect.

By following Paul's words you will be transforming the wineskin of your mind into a new and flexible one that will expand to encompass new-found discoveries. You will accomplish this by growing new neural networks and dismantling old ones. This takes time; we do not plant a seed and expect a plant to grow by the next morning.

This discovery process can be enhanced by thinking outside the box in a different sense. Some of you may be familiar with the double-entendre in the song "Thinking Outside the Box" by Rick Lang. He uses the phrase to describe the box you are buried in. It is the anticipation of something beyond the here and now that has energized many towards change in the present.

Recovery from addiction requires the exploration of new subjective landscapes and the construction of a map that leads out of the desert and into a promised land.

Thematic Takeaways

Feelings: We can usually establish a link between a feeling and the condition that evoked it. An event we were anticipating is cancelled and we feel disappointed, or perhaps an examination (physical or educational) approaches and we feel apprehensive. Sometimes, however, a feeling arises for which we have no explanation. These situations bring attention to unconscious processes and shine a light on our motivations and dysfunctions.

Needs: Understanding is a primary need. The needs list at baynvc.org/list-of-needs/ groups the following needs under the umbrella term "understanding": awareness, clarity, discovery, learning, making sense of life, stimulation. When these needs are unmet, a host of negative feelings arise which the individual may seek to escape through their addictions. Making sense of life (a primary objective of this book) provides the clarity we need to see where we are coming from and where we are going. This relieves the tension underlying many of our negative emotions.

Neural Activities: Thought is an activity we can use to explore aspects of reality that have yet to be observed. The thought experiment identifies landmarks that help us traverse the landscape of compassionate curiosity. In exercising curiosity, we gain understanding into the "why" behind our feelings and the actions they promote. Thoughts are higher-order functions that also generate feelings. Learning how to control the feelings that arise from our thoughts is the purview of cognitive behavioral therapy (CBT,) a commonly used tool in secular approaches to addiction recovery.

Relationship: Relationships with our fellows exist in the lower-left quadrant of the AQAL matrix, the inter-subjective. This is the perspective of reality from which we speak in "we" terms. We also have a relationship with ourselves in the upper-left quadrant. Lower-left conflicts arise due to differences in the world views held by individuals, but conflicting worldviews can also exist within a single individual. This condition arises when trauma that occurs at one level of psychological development is carried over into another. In response to the trauma, we latch onto a strategy for managing it at our current level of development (usually in childhood or adolescence.) We choose this strategy using the values we hold at that level, after which it becomes part of our identity. As we advance through different worldviews, the old strategies we have internalized clash with our new value systems, and this causes mental turbulence. We also begin to deny the existence of those more primitive parts of ourselves and project them onto others. This is how our shadow manifests, a subject I discuss in detail in chapter 15.

What you can do now: Learn how to control your thoughts, and the feelings they evoke, using Cognitive Behavioral Therapy's journaling process. While CBT counselling can be expensive, there are several affordable apps that help guide you through the process. You can find these by searching for CBT in your app store of choice.

5

The Foundations of Addiction Recovery

"Though no one can go back and make a brand new start, anyone can start from now and make a brand new ending."
—Carl Bard.

OUR SUBJECTIVE EXPERIENCE OF the human condition, while largely chemically mediated, consists of emergent states and conditions that transcend the physical. Chemistry is merely the foundation on which these states are built. This book, however, would be incomplete without discussing the production and distribution of narcotics whose actions at the physical level are so problematic. While chemical addiction is harmful to individuals and those close to them, it is in the harms to society that it has its greatest impact.

In my childhood home in 1970's England, the subject of addiction was a taboo. Addicts were abhorred and the risk of becoming one was viewed with trepidation. Drug use was regarded as a character flaw, a crime, and a surrender to hedonism. This view, which objectifies drug users and ignores the true nature of their addiction, is rooted in the lower-right quadrant of the AQAL matrix. Here legal and moral systems are spawned by society in isolation from any subjective or cultural experience.

Harm Reduction and the War on Drugs

The western world is now moving away from this perspective and towards a strategy based in harm-reduction[1].[2] The war on drugs, which was initiated by US President Richard Nixon in June 1971[3] has *increased* drug use and has raised the stakes for all parties involved. The riches that stem from artificially inflated black-market prices have made drug production an attractive career for people at all levels in the supply chain.

Trillions of dollars have been poured into law enforcement since Nixon issued his paradoxical fiat[4] and this has caused the producers to up the ante in response. The spread of violence, the possession and use of illegal weapons, and the employment of para-military tactics have damaged tens of thousands of lives. The latest scourge in some parts of the world, including Canada, comes in the form of highly potent opioid drugs like fentanyl (one hundred times as potent as morphine and twenty times as potent as heroin) and carfentanil which is one hundred times as potent as fentanyl.[5] Because of their prodigious kick, very tiny quantities of these drugs can be transformed into large batches of product. One hundred kilograms of heroin would fill the back of the average station wagon, but because carfentanil is two thousand times as potent as heroin, fifty grams (about the weight of a candy bar) will generate one hundred kilograms of product when mixed with fillers. This means the drug can be sent around the world, by regular mail, in small packages which often evade detection by customs officials.

The tragedy of this situation is seen when the raw material is poorly processed. In some cases, these powerful drugs are being mixed with fillers using coffee grinders.[6] The arbitrary nature of this process results in wide variations in potency from batch to batch, and each batch contains hot spots where the drug was not broken into small enough fragments, or the fragments were not evenly distributed. Users are never sure if their next fix will be their last, but the craving for the drug and the fear of withdrawal override their natural survival instincts.

1. Hyshka et al., "Harm Reduction."
2. NHRC, "Principles of Harm Reduction."
3. Dholakia, "Fifty Years Ago."
4. Murray, "Drug Abuse Peaks."
5. PubChem, "Carfentanil."
6. Forbes Magazine, "Inside."

Today's Toxic Drug Supply

I recently received a letter from the administration at my daughter's High School informing us that a seventeen year old student had passed away the previous day. We hoped that this was a natural death, but the next day the newspapers confirmed our fears. She died after using fentanyl-laced cocaine. She was not aware there was any risk as she always paid extra money to obtain the drug from sources she trusted. Sadly, trustworthiness in that business is rare. Fentanyl has made its way into every illicit substance available, making it impossible for even the dealers to know the impact of what they are selling.

On April fourteenth, 2016, the Opioid overdose crisis was declared a public health emergency in British Columbia.[7] Thousands of BC residents die each year from Opioid overdoses[8] and some of the stories are heart-wrenching. One couple, who had just occupied a new rental property with their two-year-old toddler, bought something to help them celebrate. By the next morning, they had both died from fentanyl-induced respiratory failure and their toddler was an orphan.[9] This type of situation is directly influenced by market forces and the need for traffickers to evade detection by law enforcement agencies. It is a glaring testament to the failure of the war on drugs.

A Lower-Right View of Drug Users

People who view this from the lower-right quadrant might play a script in their heads that denigrates the addict, but this ignores the fact that many of the deceased are casual drug users and not addicts. These are contributors to society whose loss hurts more than just those close to them. The situation has been at crisis proportions for years and even those with the lower-right perspective are realizing that their position is irrational. Throwing money and resources at the problem has had little effect.

The lower-right perspective is not always negative, however, in fact it is both a necessary and unavoidable feature of human consciousness. The problem occurs when the other quadrants are ignored. In that case the system, a manifestation of lower-right thinking, becomes all-important and the individual is seen as dispensable. Rather than engaging the

---

7. Provincial Health Officer, "Provincial."
8. Public Safety and Solicitor General, "More than 1,600 lives."
9. Burritt, "Hardy and Amelia."

upper-left principles of compassion and empathy, the cold, hard glare of the lower-right is directed at the problem and strategies are determined using objective reasoning alone.

This approach is common at times of war, most notably in the Germany of the mid-twentieth century. Consider this quote from the movie *Downfall* (2004) for example. Although Hitler may not have spoken these exact words, they certainly summarize aspects of his philosophy:

> Life never forgives weakness. This so-called humanity. . .is just priests' drivel. Compassion is a primal sin. Compassion for the weak is a betrayal of nature. (Source IMDB)

Hitler's Aryan ideology, centered on a Nordic master race, was based on the notion that social status is genetically inherited. This view was a distorted interpretation of Darwin's "survival of the fittest." This is what Hitler inferred by the word "nature" in the quote above. To have compassion for the weak would be a betrayal of a heartless, faceless nature whose modus operandi is that of natural selection. A closer look at nature, however, reveals a cooperative rather than merely competitive system.[10]

I suggest that this kind of compassionless reasoning, as demonstrated in Nixon's declaration of war, is at the heart of the current drug crisis. Nowhere is this illustrated better than by the government of Rodrigo Duterte of the Philippines whose war on drugs resulted in thousands of deaths during his tenure from 2016 to 2022. Amnesty International has described his government's actions as "Crimes against Humanity." Duterte's stance stems from the objectification of users and producers, but his simplistic "common sense" approach failed to address the complexity of the situation.

Portugal: A Case Study

Other jurisdictions have followed far more effective strategies. Portugal experienced an opioid crisis in the early 2000's that was similar to the contemporary situation in British Columbia, albeit free from the ravages of fentanyl. At its peak, 1 percent of the country's population were addicted to heroin. While the sight of needles protruding from people's arms was common on the streets of Lisbon in the year 2000, the possession of small quantities of illegal drugs was decriminalized in 2001, and now drug use is restricted to much more secluded areas. Portugal

---

10. Favini, "What if."

currently has one of the lowest drug related death rates in the world, with only sixteen overdose deaths (in a population of 10.5 million) in 2012. By 2019 this had risen to seventy one deaths[11], which is still low in comparison with British Columbia (population 4.6 million,) where 171 people died from drug related overdoses in September 2022 alone.

While decriminalization aims to help drug users in their recovery by removing the stigma associated with their habit, it is not a magic bullet. You cannot just decriminalize drugs and expect the problem to go away. For this reason, Portugal implemented a comprehensive all-quadrant solution. Here are some of the highlights:

1. The criminal code (lower-right) was amended so that users can legally carry a ten-day supply.

2. If found in the possession of drugs, a user is asked to present themselves to an administrative body which can impose fines or other penalties, but whose primary purpose is to assess what kind of user they are (lower-right.)

3. If the individual is regarded as a casual user, the commission will still consider their background to see if there are any conditions in their life—social, family (lower-left) or psychological (upper-left) that along with drug use could lead to addiction.

4. If they are judged to be dependent on drugs, they are offered a comprehensive rehabilitation program free of charge (lower-right.) This would consist of tests to assess the general health and any physiological damage resulting from drug abuse (upper-right,) and psychological counselling (upper-left)

5. Prior to 2001, 90 percent of Portugal's drug budget was assigned to law enforcement and 10 percent to rehabilitation. Since decriminalization 90 percent has been assigned to rehabilitation and 10 percent to law enforcement (lower-right.)

6. Changes in the law have aimed to re-orient the culture away from a judgmental view of drug use and towards a more compassionate stance (upper-left) that treats it as a health and social issue rather than a criminal one. This has resulted in the de-stigmatization of the issue, a necessary pre-requisite for open public discussion.[12]

11. Statista, "Drug Overdose Deaths."
12. Duran et al., "Guidelines."

An Integral Approach to Recovery

John Dupuy outlines a comparable approach in his book *Integral Recovery*.[13] In his experience as a Twelve Step Program facilitator and wilderness rehabilitation coordinator he realized that Integral Theory (of which the AQAL matrix is a part) would provide a highly effective structure within which to build recovery programs. The AQAL matrix is a mind map of conscious and unconscious perspectives. Its power lies in its ability to reveal the basic underpinnings of physical existence for all entities in all contexts. Regarding drug and alcohol recovery it can both identify where the challenges are for the patient and guide the treatment process.

The upper-right includes all assessments of the patient's physical health and indicates where behavioral and dietary changes may be beneficial. The lower-right places the focus on challenges that exist within the societal context. This might include resolving deficits in the legal, financial, and health care fields. The lower-left addresses issues in the cultural context. This is the relational space where personality conflicts, family hurts, and community needs can be addressed. Lastly, the upper-left quadrant is where the personal work occurs. Although the roots of any addiction inhabit this quadrant, they cannot be resolved without addressing all the other quadrants simultaneously. Dupuy also introduces Spiral Dynamics (another component of Integral Theory) and shows how its *levels of psychological development* help tailor a recovery program to the world view of the individual being helped. We will be covering Spiral Dynamics in chapters 16 and 17.

The Role of Structure in Addiction Recovery

The importance and weight of the concept of "structure" in recovery is not easy to grasp intellectually, but it is impossible to function as a human without it. We all live within cultural, societal, physical, and self-structures, the self being the seat of consciousness. From the day we are born the nurturing we receive from our caregivers, and the things we experience, become components of the internal structure we build over time. It constitutes our image of our body, our sense of worth, and our faith in values and ideals.

If something is missing in care or experience, or some trauma occurs that damages the structure during development, it will become

---

13. Dupuy, *Integral Recovery*.

unstable through the creation of mental turbulence. We all have instabilities resulting from deficits in our formation, but drug addiction is a symptom of self-structure *disintegration*.[14] While a high-functioning individual may have a job, spouse, family, and home, for some drug addicts their entire existence revolves around getting their next fix. This involves stealing from family members or strangers to fund their habit. All bridges with others are burned and the ego is the only component of the self that is left to feed. The difference between the CEO of a major corporation (who is highly regarded in our culture) and the addict on the street (to whom little value is ascribed) is a structural one. While our self-structure is principally built by others during our formative years, as adults the task of rebuilding is chiefly ours. A physical structure that has been damaged or destroyed can be rebuilt by anyone, but a self-structure cannot be reformed without a decision of the self.

The Anatomy of Addiction

Some health care practitioners, Dr. Gabor Maté among them, find themselves exploring the anatomy of addiction rather than their patients' physical bodies for much of the time. While running a health clinic in the Downtown East Side of Vancouver for many years, Maté was exposed to the physical and psychological ravages of drug addiction daily. Based on his experience in the field, he authored the book *In the Realm of Hungry Ghosts: Close Encounters with Addiction*.[15] The book focusses on what drives addiction and describes the associated costs. He shares the stories of a community caught in cycles of reform and relapse, of love and rejection, and the constant aching of profound need. His unique perspective is made even more relevant in the light of his own addiction to buying Compact Discs. This vice may sound trivial, but he occasionally spent one thousand dollars on CDs in a single trip to the music store. This behavior strained both his finances and relationships, and he struggled with the urge constantly.

One of the key lessons he learned from his patients' stories is that each of them had traumatic experiences early in life, often involving sexual, emotional or physical abuse. These situations typically spanned multiple years. He himself began life in a dire situation. In December

---

14. Zepinic, "Disintegration."
15. Mate, *In the Realm*.

1944, when he was less than a year old, the Crossed Arrows (a close ally of the German Nazi party) forced him and his mother into the Budapest ghetto. His mother enlisted the help of a stranger who smuggled him out of the Ghetto and delivered him into the care of his aunt for the next six weeks. While this separation might be considered short, it had profound effects on his mental and emotional development. Maté uses his own experience, inner dialogue and felt-sense awareness of this trauma to build an empathic link with his patients.

While his book provides no magic cure for addiction, he concludes that a stress-free environment during a child's early years is key to avoiding the circumstances that give it form. What is required is a stable, strong, compassionate upbringing where all the physical and psychological needs of a child are met. These needs include regular physical touch and eye contact. His research also indicates that a propensity for addiction begins in the fetus. If the mother is stressed during pregnancy her stress hormones can pass across the placental membrane and into the developing child. While it may be impossible to avoid stress during the formative years of a child it is possible to create social policies that minimize the risks. Recent studies have shown that interruptions in contact with the mother in the first two years of life can have negative effects on mood and behavior by the age of three.[16] As I write, child psychologists and governments are assessing the implications of this research on the children of working parents. This is likely to inform policies such as the government sponsorship of extended-period maternal/paternal leave.

Brene Brown and the Lower Right Quadrant

While this principle may contribute to the prevalence of addiction, it does not explain the situation mentioned by Brené Brown in her viral TED Talk from 2010.[17] Brown is a shame and vulnerability researcher from The University of Houston Texas who suffered her own mental health crisis in 2006. While she does not describe her situation using the language of Integral Theory, it is clear from her presentation that she was preoccupied with lower-right thinking. This is illustrated most clearly by her rapturous embrace of the claim "If you can't measure it, it doesn't exist" and her obvious aversion to another statement made by her

---

16. Howard et al., "Early mother-child separation."
17. Pangambam, "The Power of Vulnerability."

psychologist "it isn't good or bad, it's just what it is." Her research goal was to build an objective model of shame through the analysis of people's personal stories. She allocated a year to this process but eventually spent six years attempting to impose lower-right thinking on the facets of upper-left experience.

What she discovered was that the invulnerability of the lower right was a sham. Lower-right thinking attempts to build a map of the cultural and psychological space in which each feature has a corresponding symbol. She referred to this as "analyzing, identifying and putting it in a bento box." Instead of confirming her theories, her research revealed fundamental flaws in her belief in the invulnerability of the objective. This led to what her psychologist described as a spiritual crisis, but which she humbly admits was a mental breakdown. Although she does not reveal details of her breakdown, she does admit it took months of counseling before, in her words, she "got her life back." Brown identifies the lack of vulnerable connection within our society as a driver behind our generation's status as "the most in-debt, obese, addicted, and medicated adult cohort in U.S. history." Let us take a dive into why those in the Western cultures of the early twenty first century, while not necessarily drug addicted, are at least over-medicated.

Rat Park

Journalist and scientific commentator Johann Hari[18], whose family situation was steeped in drug culture, sought answers to his family's suffering. He discovered that although much of what we believe concerning addiction is based on rat studies, only the earlier studies were considered when formulating drug policy at the national and international level. These studies consist of drug use scenarios played out on caged rats which revealed a rat will self-administer cocaine-laced water repeatedly until they overdose and die. The hypothesis resulting from these studies identified drugs as the root cause of addiction. A rat will administer a drug to bring about a rewarding response, but the drug will lead to neurophysiological and neuro-chemical changes in the brain that create dependency. Ergo, removing drugs from the equation will solve the problem (cue the war on drugs.)

---

18. Pangambam, "Everything You Think."

THE FOUNDATIONS OF ADDICTION RECOVERY

Later studies conducted by the Canadian researcher Bruce Alexander[19] revealed a more nuanced picture. He created a living environment for rats that he dubbed "Rat Park" in which all their physical, psychological and social needs were met. They had plenty of company, exercise wheels, and tunnels to explore, and access to plain water and cocaine-laced water. In that environment the cocaine-laced water held no more allure than pure water. This appears to suggest that the obsession with the drug was a result of unmet social and environmental needs. It also indicates that a drug-induced reward activates the same neural pathways that process social connection. Without social connection there is a psychic vacuum and, as the cliché goes, nature abhors a vacuum. A solitary caged rat will fill that vacuum voraciously with cocaine. One in a healthy social environment will satisfy that void with the joys and trials of living in community.

Applying the Lessons of Rat Park to a Human Cohort

If this is the way rat culture operates, can this principle be applied to human cultures also? It would certainly be unethical to place a human in solitary confinement and feed them drugs, but humans have been placed in situations that are similar with comparable results to Alexander's Rat Park experiments. Studies on troops returning from the Vietnam war reveal what happens when a cohort of humans are placed in situation where their psychological needs remain unmet for an extended period.[20] Obviously, Maslow's primary need for safety (see Maslow's Hierarchy, chapter 6) remains unmet in such situations, but towards the end of the war morale also became a major issue. As it became clear that the war was un-winnable and doubts arose about the value of the enterprise in general, the troops began to self-medicate with heroin. By the end of the war 20 percent of the troops had become habitual users.

The traditional model of addiction predicts that these troops would require extensive rehabilitation and potentially life-long drug treatment following their return, but this is not what happened. Once they returned to their home country, their families, friends, and creature comforts, the majority just stopped using. It is true that approximately 10 percent of them experienced persistent cravings, and this might be an indication of

---

19. Sederer, "What Does."
20. Hall and Weier, "Lee Robins."

the genetic pre-disposition towards addiction that has been indicated by several studies.[21] Nevertheless, the remaining 90 percent of that cohort went on to lead a life of abstinence.

The Use Of Heroin in Pain Management

Again, if drugs are the cause of addiction to drugs, we would expect everyone who has prolonged exposure to heroin to become an addict. Experience in the medical setting also refutes this claim. The follow-up to surgical procedures includes pain management strategies that often involve opioid drugs. One of the most common of these is diamorphine which is heroin by another name. If heroin is what causes addiction to heroin, we would expect to see a stream of addicts emerging from hospitals all over the world, but this is not the case. But what about the long-term use of opioids in the treatment of chronic pain? Experience shows that this introduces its own set of challenges.

If humans seek out pleasure and avoid pain it is logical to conclude that a drug that helps manage pain is in and of itself addictive in its ability to relieve suffering. Add to that the obvious neuro-physiological adaptations that occur in the body because of drug use and there is ample reason to become physically dependent. The body of a healthy individual operates in a state of equilibrium where signaling molecules perform their intended functions. If this function becomes problematic, as is the case with chronic pain, the use of opioids for pain management creates a new equilibrium. If the treatment is discontinued, the physical adaptations remain and the body must adapt again, resulting in withdrawal-induced pain. All this occurs at the physical level and results in physical dependence, but not necessarily addiction.

The ISO OSI 7-Layer Model—An Analogy

We can further explore the anatomy of addiction by drawing an analogy from the field of computer science. Opioids are signaling molecules that affect the propagation of neural impulses to, from and within the brain. This might involve stimulating or inhibiting electrical activity. Neurons are analogous to the network cable that connects your computer to the internet, and the signals propagated along the neuron are

---

21. NIDA, "Genetics and Epigenetics."

like the voltage-dependent binary messages that traverse computer networks. Computer networking is formally defined using a seven-layer model in which the physical layer (the cables and network hardware) is layer one, and the presentation layer (the user interface) is layer seven. In between these two are five other layers of software that manage the routing of messages and the management of conversations between the sending and receiving parties. In the context of your web browser, the presentation layer is that beautifully formatted web page with its alluring imagery. In the human context the presentation layer is the face we present to the world.

Much of the field of psychology is concerned with the layers of conscious and unconscious processing that occur between the physical realm of neurons (the physical layer) and the behavioral domain of everyday life (the presentation layer.) In both of our examples, changes in the physical layer will affect the operation of other layers. In the computer network context, if there is interference from an external magnetic or electrical source (which we could describe as electrical turbulence) the software at levels above the physical layer must handle this accordingly. Likewise, issues at the physical level of the biological organism affect processing throughout the system.

While a computer network operates using discrete voltage differences (usually plus five volts for a binary one and minus five volts for a binary zero) the human brain has many distinct levels of "on" and "off" (spike amplitude or frequency) that are mediated by different messaging chemicals. Each signaling molecule controls the operation of a discrete system that often interacts with other systems using complex feedback loops. The face we present to the world through expressions of behavior and our own subjective states can be modified by turning the volume up or down in any of these systems individually. For example, our mood and energy level can be improved by increasing activity in the serotonin-signaling mechanism by administering an anti-depressant drug. However, due to the complexity of the brain and the interdependence that exists between its components, a situation exists that has no analogue in our computer networking example. In the brain, *issues at higher levels can affect operation at the physical level*. Treating a dysfunction at the physical level may appear to normalize function, but it is only a resolution of the symptoms and does not address the root cause. Consequently, doctors normally suggest a course of anti-depressants be accompanied by talk therapy.

While researchers are probing the physical level for a genetic explanation for addiction, in most cases it is resolved at the self-structure level. A genetic cause may explain why some individuals are unable to manage their addiction without lifelong drug treatment.

Drugs and Spiritual Practice

Two other perspectives also shed light on the subject at hand. In some cultures, drug use is a component of spiritual practice. While these same drugs may be widely abused in Western culture, this is not the case when use is governed by spiritual traditions. Another perspective involves the use of psychedelic drugs in the treatment of addiction. While substances such as LSD and psilocybin were used in medical research in the 1950s and 60s, the war on drugs curtailed this practice. Now the tide is turning, and research has resumed into the use of these substances for the long-term treatment of depression and conditions that have, up until now, been addressed using talk therapy.

Treatment with the dissociative hallucinogen ketamine has been shown to prevent the recurrence of major depression for months at a time. Micro-doses of Psilocybin, the active ingredient in magic mushrooms, have also been used to the same effect. The accepted logic is that hallucinogens do not carry a risk of addiction, but for some people this is simply not the case. Any drug that alters consciousness in a way that alleviates emotional or physical pain can be used as a crutch and this can always become habit-forming. Even the antihistamine Benadryl proves addictive to some because of its hallucinogenic anti-cholinergic side-effects when taken in high doses. As with all drug treatments, the benefits must be weighed against the risks.

A special case involves the use of Ayahuasca. This is a plant-based compound used as a traditional spiritual medicine in ceremonies among the Indigenous peoples of the Amazon basin. Its active ingredients include dimethyltryptamine (DMT) and a monoamine oxidase inhibitor which prevents the DMT from being neutralized when consumed orally. People who have consumed Ayahuasca report having spiritual revelations regarding their purpose on earth, the true nature of the universe as well as deep insight into their future mission in life. It has even been incorporated into Christian worship by sects such as the Brazilian Santo Daime whose first European-affiliate churches sprung up in the Netherlands in the late twentieth century.

The use of psychedelics is believed to provide access to aspects of consciousness that are normally beyond reach. From these altered perspectives, an individual may obtain a route to discovering previously unknown truths which can affect the resolution and healing of past trauma. I will not be focusing on psychedelics as their effects on the brain, while they might lead to alluring states of being and could become habit forming, do not fit the profile of most addictive substances (they do not primarily act on the limbic system.) But before we conclude this chapter, a word about the other major contributor to addiction and drug overdose deaths.

The Role of Big Pharma in the Opioid Crisis

Opioid pain killers are a vital tool for reducing the suffering of millions who, following injury or surgical procedures, experience moderate to severe pain. Pain can also assume a chronic aspect where the cause might not be immediately obvious. This type of pain also responds to opioids. The ethereal nature of chronic pain is illustrated by Phantom Limb pain, a condition where the patient feels discomfort in a part of the body that has been surgically removed.[22] This too can be treated using opioid drugs. Chronic pain is particularly problematic, as the long-term administration of opioids will always bring about physical adaptations in the body's opioid receptor system which lead to physical dependence (but not necessarily addiction.) For many, opioids also produce marked limbic effects (feelings of wellbeing) which add a psychological component to their use.

While the judicious use of such drugs may be of great benefit, the pharmaceutical industry has on occasion, overstepped its mandate and pursued profit over the health of the general population. The American pharmaceutical company Purdue Pharma is a case in point. Purdue followed a marketing strategy that progressively increased the dosage guidelines for their flagship opioid-based product, Oxycontin, for no valid reason. They also lied about the addiction potential of the drug and promoted it as a less-addictive alternative to its competitors. In 2020, after a series of lawsuits that stretched over a period of almost two decades, the company reached an agreement to pay $8.3 billion in compensation for harms incurred. In the words of the settlement, they had:

---

22. Dingman, *Bizarre*, 44–45.

"Knowingly and intentionally conspired and agreed with others to aid and abet" doctors dispensing medication "without a legitimate medical purpose."[23]

Some patients used Oxycontin to treat pain that stemmed from an observable cause, others used it to numb their trauma-based psychological wounds and still others used it to treat trauma-related chronic pain. The cause of such pain cannot be seen objectively, but it can still be agonizing.[24] Purdue viewed this cohort as a stable revenue source and targeted them relentlessly through the legal, public relations, and policy routes. This led to the overprescription of Oxycontin and thousands of unnecessary deaths. The health of many survivors was also permanently compromised through overexposure to the drug.

The dual scourges of the illegal drug trade and the unethical pharmaceutical company drive most drug-related harms to society, but they are both motivated by the same underlying principle—the love of money. Chapter 8 deals with this in detail, but before we can grasp the allure of money in objective terms, we must first explore the primary mechanism underlying psychological addiction. This we will do in the next chapter.

Thematic Takeaways

Needs: Our primary need is to maintain dopamine flow, a process that is driven by needs satiating behaviors of all kinds. Isolated rats might accomplish this by drinking cocaine laced water, humans trapped in low morale (low dopamine) situations might turn to heroin. But healthy social and cultural environments, and the self-structures they create, are the forces that drive *ambient* dopamine flow by meeting love, belonging, and esteem needs.

Neural Activities: Although amphetamine, through its action on the TAAR1 receptor, is the closest drug-based analogue for relationship, all drugs that promote dopamine activity create similar phenomena, this includes cocaine and the opioids.

Relationship: Relationship is the primary arena in which addiction operates. Often a strategy for coping with abuse, addictions feed on the self-blame, self-loathing, and shame that stems from the actions and words

---

23. Wikipedia Contributors, "Purdue Pharma." para. 4.
24. Gasperi et al., "Pain and Trauma."

of others. These experiences and the behaviors they engender taint all future inter-personal relationships, making addiction a family condition rather than an individual one. In the absence of abuse and trauma, an individual might still turn to drugs to maintain ambient dopamine flow. A healthy social/relational environment satisfies this need.

What you can do now: Familiarize yourself with the list of feelings at baynvc.org/list-of-feelings/ and explore how these show up when your needs are met or not met.

Part Two—*Understanding:*
Why Addiction Takes Hold

6

Dopamine

"A deep sense of love and belonging is an irreducible need of all people. We are biologically, cognitively, physically, and spiritually wired to love, to be loved, and to belong. When those needs are not met, we don't function as we were meant to. We break. We fall apart. We numb. We ache. We hurt others. We get sick." —Brené Brown

Modern Neuroscience tends to use a networking model for how the brain works. From this perspective the brain is composed of separate devices linked together by lines of communication. Each device has a specific function, but the complexity of their interactions gives rise to behaviors that emerge from their joint influence. In the past this model attempted to ascribe each behavior to a specific brain region, but as our understanding of the brain progresses it becomes clear there is no unitary seat for each complex operation. A complex behavior only comes into being through the interaction of multiple domains and the structures that connect them.[1]

The difficulty for us is in determining the boundary between simple and complex. There are some functions that are local to one region. These are normally brought to light by the effects of injury or dysfunction on the operation of that locale. For example, an injury to Wernicke's Area in the left temporal lobe can affect the individual's perception of written or spoken language, and injuries to the Hippocampus can prevent

1. Dingman, *Bizarre*, 45

the formation of new memories. But it would be inaccurate to assume those areas are the only ones involved in those functions. This distinction becomes more apparent in functions such as the appreciation of music and the exercise of artistic expression. There is no *nucleus musicalis* that processes the chromatic structure of sound; neither is there a *sculptus oblongata* that guides the sculptor's hammer and chisel. Both behaviors result from activities that are distributed across the brain.[2]

The Limbic System as a Distinct Entity

While there is no single brain region that gives rise to addiction we can still speak in terms of localized function and how that might contribute, as well as how the connections between different locales are involved. We can also draw on our knowledge of structure to build metaphors which, while they may be abstractions, provide useful and relevant talking points. For example, thinking of the limbic system as a distinct entity is more of a convention than a scientific reality. Yet the concept, which pervades nearly all research on addiction, is a thinking cap that is worn by many in the field. As addiction is behavioral in nature, an understanding of the processes and structures underlying motivation and reward is key. For that reason, this book will discuss the nucleus accumbens and amygdala in some detail. The nucleus accumbens for its role in reward and the amygdala for how it contributes to aversion.

What Motivates Us?

Our behavior has a major effect on those around us, and for that reason it must be moderated for the benefit of all, but what drives our behavior? Are we the one in control or are influences at work that transcend our conscious involvement? We have seen that the limbic reward pathway innervates the enjoyment that flows from a tasty meal, a good glass of wine, the thrill of a loving touch, and the beauty of a sunset. We also appreciate that because these things are enjoyable, we pursue them. These drives animate our everyday behaviors on a deep level. We could describe them as selfish, but we also enjoy being selfless—as the adage goes, "it is better to give than to receive" (Acts 20:35.) While the primary function of limbic physiology is to ensure the survival of the individual and to reward key

---

2. Sacks, *Musicophilia*, preface.

behaviors to that end, it also contains the seeds of its own demise. The overconsumption of food, while enjoyable, has profoundly negative effects on many bodily systems; an overindulgence in the bottle destroys the liver, and a hyperactive libido promotes the spread of disease.

Needs and Wants

The *raison d'ê·tre* underlying each of these drives is the procurement of a specific physical or psychic resource that promotes health in the mind/body system, but each of these drives can exceed their mandate, resulting in disease or dysfunction. Eating to live is one thing, living to eat is another. In the 1950's, Abraham Maslow identified a collection of primary motivations shared by all humans which he referred to as *needs*.[3] Other researchers working in this space have used the word *wants* as an alternative. Years ago, I overheard a conversation on the beach that helps illustrate the distinction between the two. A child desired a shovel his sister was using to build a sandcastle. There was only one, and this scarcity introduced some tension into the situation. The little boy repeatedly stated, "I need it," then after concluding there would be no response from his mother he said with an indignant tone "I want it." Did he need the shovel or want it?

---

3. McLeod, "Maslow's Hierarchy."

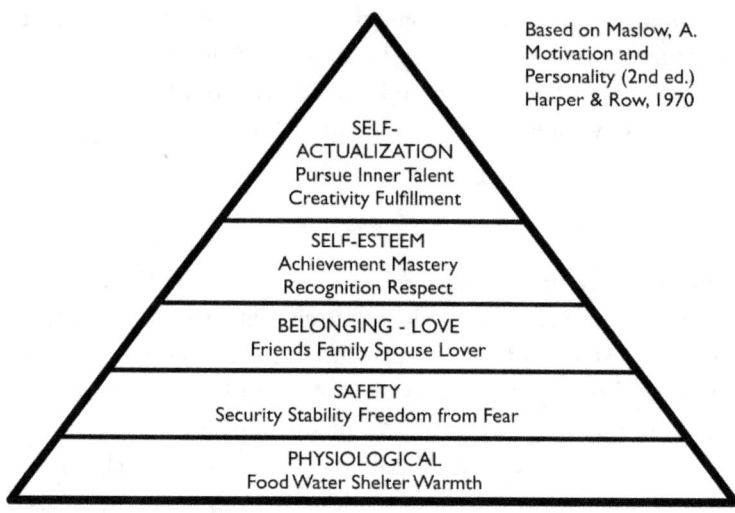

Figure 6.1 Maslow's Hierarchy of Needs. Wikimedia Commons.

Our culture sees the act of *wanting* something as a trifling with the frivolous, while *needing* something is an urgent pining for the essential, but this is not necessarily the case if needs are viewed through Maslow's lens. His hierarchy begins at the lowest level with a set of physical inputs that the human body requires to survive. These include food, water, and air. A human can exist without air for a few minutes, without water for a few days, and without food for a few weeks, but after those respective periods of time the result is always the same—death. These inputs can surely be called *needs* as we need them to survive. The next level of the hierarchy includes safety and security. Here things are a little less concrete. We can survive without safety for a long time, but the chances of survival are lower. If we live in an environment full of predatory animals there is a higher probability that our life will be taken early, but many people in such an environment live to a good age. Survival at this level is no longer contingent on absolutes but rather probabilities.

Do we *want* security or *need* it? The child on the beach could have built a sandcastle with his bare hands, but this would have taken longer than by using the spade. He *needed* the spade to complete his task with a measure of ease. In the same way security provides a level of ease that forms the foundation for contentment. Without the risk of attack by wild animals there is no need to set a watch or build secure accommodations.

In this case security is *needed* to provide the level of safety and ease to focus on our relationship with ourselves and others.

Belonging and Love

The next level of the hierarchy involves belonging and love. The child on the beach would see the gifting of the spade as a sign that his mother loved him and that withholding it might introduce doubts about that. There is also the issue of sibling rivalry and how important the boy felt in the family context. This situation is at Maslow's level of esteem needs. All these non-essential factors could be described as wants, but if viewed as pre-requisites for a particular goal they become needs. If the goal is to feel secure in our family relationships, then the psychic resources of love, appreciation and respect are vital contributions towards that end. Put another way, if the ceilings in our house are twelve feet high, we would *need* a ladder to change the light bulb. It is not that we need a ladder for survival, humans survived for millennia before the ladder was invented, but a ladder is required to reach the goal of a newly installed light bulb. What we tend to absorb through a process of cultural osmosis is that survival needs are the only real needs, a perspective that is both limited and pernicious. If our goal is to be fully functional human beings then safety, love, belonging, and self-esteem are vital pre-requisites.

The Parable of the Empty House

The gravitas of need is illustrated in the parable of the empty house recounted in Luke's account of Jesus' words in Luke 11:24–26 (ESV):

> When the unclean spirit has gone out of a person, it passes through waterless places seeking rest, and finding none it says, 'I will return to my house from which I came.' And when it comes, it finds the house swept and put in order. Then it goes and brings seven other spirits more evil than itself, and they enter and dwell there. And the last state of that person is worse than the first.

Nature abhors a vacuum and needs create a vacuous state that fills itself with whatever can be obtained, either good or bad. The top three tiers of Maslow's hierarchy (social belongingness, esteem, and self-actualization) are in the relationship and connection space. The human mind cannot grow into the mold of its potential without nurturing, mentoring, and peer relationships. While the parable directly addresses the state of a repentant adult whose negative life strategies have been driven out but not replaced with productive ones, the same principle applies to the nascent identity of the child. Without a sense of belonging in the family, third-level needs will be met in the peer group, potentially in a gang environment. In this case fourth-level needs will be attained by pursuing the goals of the gang and acts of theft; destruction and violence will be used to gain esteem and status. Fifth-level needs could be fulfilled by ascending the gang's organizational hierarchy.

If the relational influences experienced during child development fail to fill the vacuum with positive determinants, the identity becomes malformed. The features of this identity no longer jibe with the attachment points of healthy role models, the places where connection and emotional refueling take place, driving the individual towards other means of dopamine regulation such as substance abuse and narcissism. The incentive to meet needs is overpowering, but they can be met in a constructive or destructive manner. All too often belonging, esteem, and self-actualization needs become a matter of life and death—when they go unmet and suicide becomes the ultimate solution to our problems.

Attachment Styles

Healthy adult influences during childhood are vital for the development of functional attachment points. Four different attachment styles have been identified in children[4]:

> *Secure attachment* occurs when children feel they can rely on their caregivers to attend to their needs of proximity, emotional support, and protection.
>
> *Anxious attachment* occurs when the infant feels separation anxiety when separated from the caregiver and does not feel reassured when the caregiver returns to the infant.
>
> *Avoidant attachment* occurs when the infant avoids the caregiver.
>
> *Disordered attachment* occurs when attachment opportunities are unpredictable.

---

4. The Attachment Project, "Attachment Styles."

Secure attachment is the most advantageous and only healthy mode of attachment. In a sense these styles present a chicken and egg situation, as the style adopted by the child will result from their previous experiences with attachment. The last three styles involve an overactive amygdala, a brain region that plays the role of protector of the *self*. This region becomes activated when external support and protection are lacking. The influence of the amygdala also grows with use, creating a kind of domino effect. Children who were separated from their parents during the first two years of life, who were adopted or insecurely raised by multiple caregivers will have difficulty meeting their own attachment needs later in life. This underscores the importance of relationship in building the self-structures that give form to a healthy adult.

The Role of Dopamine in the Hierarchy of Needs

So, what is the role of dopamine and the incentive/reward pathway in all of this? The purpose of dopamine, the fuel on which the nucleus accumbens runs, is to drive the individual to meet their needs at all levels of Maslow's hierarchy. This is done by rewarding behaviors that feed the self, either physically or mentally, with feelings of wellbeing. These actions include eating, socializing, overcoming challenges, exercising, relaxing, creating, loving, and learning. These good feelings can be described as *feelings experienced when needs are met*. If an individual lacks the opportunity or ability to experience these feelings by performing life-serving actions, they will use destructive strategies instead.

The meeting of needs becomes important due to a second set of feelings, *feelings experienced when needs are not met*. These feelings are generated, for the most part, by the amygdala (more about that in chapter 14.) These feelings are unpleasant and include anger, frustration, nervousness, irritability, concern, and anxiety. Addiction is the attachment to destructive strategies that avoid the feelings flowing from unmet needs and produce feelings that normally arise from meeting needs. Once a strategy has been identified that avoids unpleasant feelings, the resources needed become irrelevant (for example *love*) and maintaining a state of wellbeing using the replacement strategy assumes prime importance. A chemical or behavioral strategy that promotes high levels of activity in the nucleus accumbens will temporarily overwhelm any unpleasant feelings stemming from unmet needs.

Anticipation and Reward

The nucleus accumbens provides the incentive to act and the incentive is created by an expectation of reward. The reward is a good feeling, be it an enjoyable sense of flavor, joy, pride, thankfulness, awe, optimism, or relief. When we *anticipate* a reward, the dopamine levels in our nucleus accumbens spike and this drives us to act. We take an action that meets a need, and the dopamine levels rise again to provide the reward for our actions. The sensation of reward is multi-faceted. It involves dopamine for its expression, but the subjective experience of the reward can take many forms.

Think of the nucleus accumbens as a light bulb that can display any color imaginable. We might experience one rewarding sensation from the scent of Chanel Number 5, another while watching a Lamborghini pass by and a third when savoring a bite of our favorite cheese. These sensations all take place because of nucleus accumbens activity, as electrical signals are propagated in myriad ways depending on the stimulus and our memory associations with that stimulus. The term often used to describe this is a "lighting up" of that part of the brain on a brain scan, most commonly a Positron Emission Tomography (PET) or Functional Magnetic Resonance Imaging (fMRI) scan.

We are attracted to scents, hyper cars, and foods because they light up our nucleus accumbens. The attraction we experience is not a magnetism that exudes from the object itself, but rather a pushing of our attention towards the object that can be described as "desire." Desire is life-serving when it drives us to fulfill our needs, but it can also lead to harms of the self or others when it escalates to the level of greed or covetousness, as is often the case with addiction.

Relational Attraction and the Nucleus Accumbens

Each Friday night from 2018 to 2020 I attended weekly Twelve Step Program meetings run by Celebrate Recovery (www.celebraterecovery.com.) These meetings were curtailed (temporarily) due to Covid-19 in early 2020, but, thankfully, I was able to complete my program prior to the onset of the pandemic. At the end of each session, those attending were dispersed to one of two sharing circles, two for men and two for women. Each gender had the choice of attending either the substance-use or relationship group. It was telling that the relationship group for men was usually twice the size of the substance-use group. This indicates that a high proportion of these men recognized the addictive tendencies they had exhibited in the relational sphere. While claiming that the problems shared in the relationship group flowed from desire alone would be an oversimplification, but desire was a key factor in each case.

The generalized role of the nucleus accumbens in needs-satiating behavior shows itself in the way attraction is discussed metaphorically in colloquial English. Members of the opposite sex are often described using language normally reserved for the senses of taste, smell, touch, sight, and sound. (Recent research ascribes these feelings to opioid rather than dopamine activity in the nucleus accumbens.) Consider these examples:

Sight: "He's a sight for sore eyes."
(Meaning he is very attractive to look at.)

"She's a real feast for the eyes."
(Implying she is visually stunning.)

Taste: "Look at that guy—what a dish"! or
"I met this amazing woman last night—she was delicious"!
(Extending the company of an alluring presence into the realm of taste.)

"He's like a fine wine, only getting better with age."
(Comparing attractiveness to the refinement of wine.)

"She's like a sweet treat, irresistible."
(Using the idea of a delightful dessert to convey attractiveness.)

Smell: "He's like a breath of fresh air."
(Suggesting that he is refreshing and attractive in a unique way.)

"She's as enticing as a fragrant flower."
(Comparing her attractiveness to the allure of a beautiful scent.)

Touch/Feeling: "Being with him is like a warm embrace."
(Conveying that his presence is comforting and attractive.)

"She's as soft as silk, both in appearance and personality."
(Comparing softness and gentleness to attractiveness.)

Hearing: "His voice is music to my ears."
(Implying that his speaking or singing is attractive.)

> "She has a laugh that's contagious."
> (Highlighting the attractiveness of her laughter and its effect on others.)

Setting aside the allegorical message presented by King Solomon in the Song of Songs for the moment, the poem also illustrates Solomon's familiarity with the unifying role the nucleus accumbens plays in the desire to meet needs. Consider the passages below that portray the beauty of the bride and the beloved in terms of nourishment, fragrance, the allure of nature's bounty and the satisfaction that comes from eating and drinking—all used to increase the work's literary impact (all verses from the NIV):

> 1:3 Pleasing is the fragrance of your perfumes; your name is like perfume poured out. No wonder the young women love you!
>
> 1:14 My beloved is to me a cluster of henna blossoms from the vineyards of En Gedi.
>
> 2:3 Like an apple tree among the trees of the forest is my beloved among the young men. I delight to sit in his shade, and his fruit is sweet to my taste.
>
> 4:11 Your lips drop sweetness as the honeycomb, my bride; milk and honey are under your tongue.
>
> 5:1 Eat, friends, and drink; drink your fill of love.
>
> 7:2 Your navel is a rounded goblet that never lacks blended wine. Your waist is a mound of wheat encircled by lilies.

The opening stanza of the poem, reprised in the 1985 Hit "Kiss Me" by Stephen Tin Tin Duffy, emphasizes the intoxicating nature of love:

> 1:2 Let him kiss me with the kisses of his mouth—for your love is more delightful than wine.

There would be no comparison between love and wine were there not a similarity in their rewarding effects. According to this verse, whatever delight it is that the beloved's love contributes to the bride's state of consciousness; it must exceed that bestowed by wine. On the physical level, it is the action of endorphins (endogenous opioids) that weaves love and wine together, often as companions but at times as rivals.

An Upper-Right Quadrant Consideration of Reward

Returning to an upper-right quadrant consideration of these phenomena, let us dig a little deeper into the technologies used to measure them objectively. PET scans can show the distribution of a radioactive marker drug as it binds with a target receptor or receptors. In contrast, fMRI scans show which parts of the brain are active based on blood flow. In both these cases the view of the brain is monochrome, with higher signal intensity indicating increased activity. The level of activity can be color-coded, but this is a false color view that has little relation to the subjective experience of the activity. In the following sentences I am using color as a metaphor for what the subject is feeling.

We might leave our house on a day in early March to be greeted by the smell of freshly cut grass. This could light up our nucleus accumbens with one color as we feel a sense of possibility, of hope for the arrival of springtime and warmer weather. We might also pass by the window of a department store and see a gorgeous dress in the display, our nucleus accumbens lights up another color. This rewarding feeling is completely different from the first, but it is still a result of the flash of electrical activity. In both cases the illumination of the nucleus accumbens is triggered by needs that are met—the needs for hope and beauty, respectively.[5]

The PET scan image shown in Figure 6.2 illustrates the singular role the nucleus accumbens plays in addictions to both cocaine and food. The image portrays the nucleus accumbens of healthy, cocaine-addicted, and obese individuals. In this monochrome image, the dark ovals at the midpoint of each hemisphere, in the leftmost of the three scans, indicate normal function. In the cocaine and obese examples these areas a much less defined. While this type of image appears frequently online, they are usually shown without a guide to their interpretation. To the untrained eye this image consists of three mysterious blobs. I want you to have a more comprehensive grasp of the figure, so despite the potential for being verbose, here is a layperson's description of PET imaging and what the figure is showing us.[6]

---

5. Bay NVC, "List of Needs."
6. Wikipedia Contributors, "Positron emission tomography."

Positron Emission Tomography

PET scanners, like CT and MRI scanners, produce cross-sectional images of the human body. PET scanners are the stuff of science fiction in that they employ the same process that powers the Star Ship Enterprise (matter/anti-matter annihilation.) They are commonly used in the study of dopamine-mediated function and can produce images in concert with a radioactive dopamine antagonist such as Carbon-11 Raclopride. Raclopride is commonly used in medical practice as an antipsychotic (it is a dopamine D2 receptor antagonist.) It can, however, be adapted for the PET scanning process by replacing the naturally occurring carbon atoms within the Raclopride molecule with those created in a cyclotron. These atoms have a mass number of eleven and are mildly radioactive. Carbon-11 has a half-life of twenty minutes, and when each atom decays to Boron-11 it releases a positron (the anti-matter equivalent of an electron.) After release (usually within five millimeters of the source) the positron will encounter an electron. At this point matter/anti-matter annihilation takes place releasing two gamma rays that travel in opposite directions.

The detector of the PET scanner forms a ring around the part of the patient's body being assessed, in our case the brain. If a gamma ray passes through the detector, it creates a flash of light. The detector senses this flash and takes note of the time it occurred. If two flashes happen at approximately the same time, and on opposite sides of the detector, the software assumes that these are the result of a single Carbon-11 decay. From the precise locations of the two flashes, it can deduce where in the brain this decay occurred, and from that it can infer that a molecule of Raclopride, and the dopamine receptor to which it is attached, were at that location. The amount of radiation measured at a particular locale indicates how many Raclopride molecules have bound with dopamine receptors at that point in space. These receptors were unoccupied prior to the arrival of the Raclopride molecule. A lower signal intensity in a particular area indicates that there are fewer dopamine receptors at that locale. This is clear from the image in that most of the brain shows very little signal, indicating that the vast majority of D2 receptors are in the nucleus accumbens.

Let's use an analogy here to illustrate how a PET image is constructed. The colors we choose for our example are irrelevant, but I will use black and white to correlate with the monochrome image shown figure

6.2. Imagine you have a white piece of construction paper with ten spots of glue on it. The glue spots represent the dopamine receptors. The glue is clear so the white color from the paper shows through. This is what the limbic system looks like to the PET scanner prior to introducing the Raclopride into the blood stream; it is blank. Now we use a fan to blow hundreds of black feathers over the surface of the construction paper. This represents the arrival of the Raclopride molecules as they are carried past in the blood stream. If a feather meets a glue spot it will stick. After that no other feathers can stick to that location because that glue spot is occupied. We could now determine how many glue spots (receptors) there were by counting the number of black feathers on the card. This is the equivalent of the dopamine receptor count. What we end up with is a sheet of paper peppered with highly visible black feathers.

As a practical illustration of this analogy, the PET image below illustrates how cocaine addiction and obesity are related in terms of dopamine receptor density. The pale areas in the center of each brain hemisphere correspond with the left and right limbic regions. The normal image exhibits high density Raclopride occupancy at each nucleus accumbens, signified here by the dark ovals, showing that many dopamine receptors were available for binding. In this case the dopamine receptors and the black feathers to which they are metaphorically attached (the Raclopride molecules) are infinitesimally small. In both other cases dopamine receptor density is greatly reduced. This is an adaptation the body uses to maintain equilibrium in the mind/body system. Dopamine receptors in these neurons are permanently removed or temporarily drawn to the interior side of the cell membrane.[7] The reduction of the exterior receptor count means there are fewer targets for dopamine to bind with, so even elevated levels of synaptic dopamine will have less effect on signaling.

This situation calls for a much larger stimulus to produce a normal response and will affect the subject's ability to experience satiation. This naturally drives sensation seeking behavior and over-consumption.[8] Considering the Rolling Stones' drug-fueled lifestyle in the 1960's this may well be the state of mind described in the song "I can't get no satisfaction."

---

7. Nader et al., "Effects of Cocaine."
8. Collins, "Imaging Willpower."

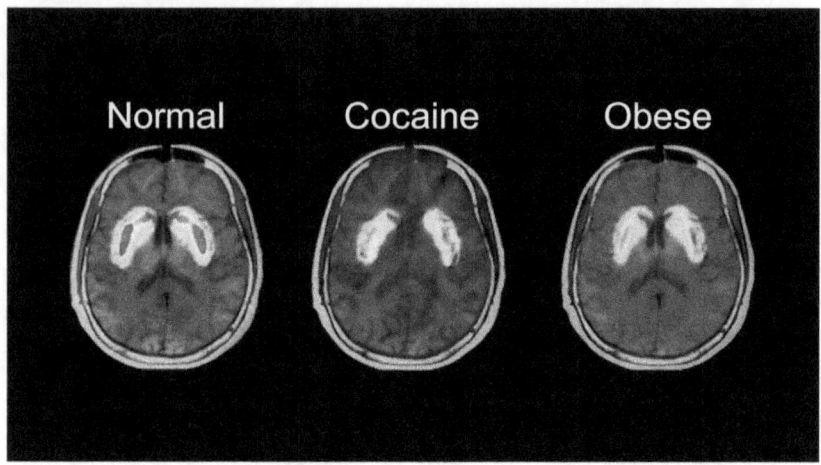

Figure 6.2 PET Scans of Three Brains. Image source: National Institute on Drug Abuse (NIDA), U.S. Department of Health and Human Services. Adapted from research by Volkow et al. Used under public-domain educational permission.

Dopamine and Aesthetic Appreciation

We can also use alcohol and other drugs to illustrate the compounding effect of mixing external and internal dopamine stimuli. One of the most powerfully rewarding dopamine responses is that of aesthetic appreciation. Color, form, art, music, and the wonders of nature all stimulate a pleasure response. When I was first prescribed Ritalin for ADD I visited a friend's house where a song was playing on the radio. Halfway through listening to the song, and enjoying it immensely, I had a realization—"I hate this song, so why am I enjoying it now"? The Ritalin had up-regulated dopamine activity causing the small amount of dopamine released into the synapses by the song to create an elevated state of wellbeing.

Alcohol can enhance our perception of beauty in the same way. While the graphic to which I refer is a joke, and some might consider it to be in poor taste, it does illustrate the principle rather well. Try googling "before 6 beers after 6 beers" to see the popular ambigram image that illustrates this principle rather well.

Oxytocin and the Endorphins

Turning back to the subjective experience of limbic stimulation, we find that dopamine does not work alone. Other signaling chemicals also bind with receptors in the nucleus accumbens and affect its firing patterns. Two drugs that follow this course of action (both of which are created naturally within the body) are Oxytocin and a group of Morphine-like chemicals called endorphins. Oxytocin modulates nucleus accumbens function directly, and endorphins influence dopamine signaling through their action on interneurons and hedonic hotspots in the nucleus accumbens shell.[9] On the subjective level the oxytocin experience is like that of a dopamine inducer[10], a response that is exhibited in both sexes. In fact, the administration of oxytocin has been shown to diminish drug-seeking behavior in methamphetamine addicts.[11] It is responsible (in conjunction with the opioids and phenethylamines) for the bliss of a warm hug and the coziness of a cuddle with the one you love. Oxytocin also promotes social connection by decreasing feelings of suspicion and engendering feelings of trust in social situations, even when these feelings may not be warranted.[12] It is also central to the formation of life-long pair bonds, as shown in the lives of monogamous Prairie Voles.[13]

Research has revealed that oxytocin, the phenethylamines, the endorphins, and by extension dopamine, express many aspects of relationship. We are hard-wired for relationship with others. Whether we find someone physically attractive or love them as a friend or family member, the activities of the nucleus accumbens are at the root of each.

Some people find this perspective offensive because it paints relationship as deterministic in nature. If relationship is merely a chemical reaction, one might think, then we act as mere automatons and have no agency over our life conditions—our lives unfold due to chemical interactions over which we have no control. This is not my view at all and my reason for presenting the activities of spirit in this way is actually the reverse. My goal is to show that our attachment to addictive chemicals stems from their role as proxies for relationship. By exercising free will in seeking out relationships (with ourselves and others) that meet our belonging and

---

9. Palm and Nylander, "Endorphin—an overview."
10. Love, "Oxytocin."
11. Edinoff et al., "Oxytocin."
12. Israel et al., "Oxytocin Decreases."
13. Báez-Mendoza and Schultz, "The Role."

esteem needs in healthy ways, we will (after a period of adaptation) have no further need for those destructive strategies.

Having read this far, I hope you have developed some awareness of what is occurring at the physical level when a craving arises and have begun to tune in to the internal sensations that accompany it. The next chapter will build on this foundation and lay the groundwork for the approaches to recovery that follow.

Thematic Takeaways

Feelings: The nucleus accumbens drives us to meet needs and rewards us for so doing with the good feelings that arise when needs are met. The amygdala produces feelings of aversion that inform our conscious minds when needs are not met.

Needs: Needs are the resources required for our minds and bodies to operate at full capacity. These may be as concrete as the need for water and air, or as ethereal as our need for love and belonging. At times, our desire for love may assume the same urgency as that produced by a voracious thirst.

Neural Activities: While we cannot correlate each complex human behavior with a discrete brain region, we can make a general claim that the nucleus accumbens is of central importance in motivating us to meet our needs and in rewarding us for doing so. The activity of dopamine receptors is central to this process. Driving high levels of dopamine activity by engaging in chemical or behavioral addictions brings about adaptations in the nucleus accumbens' dopamine system which exacerbate the situation.

Relationship: The adoption of a secure attachment style by every infant is a vital step in healthy psychological development. For many, addictions are the unhealthy means by which the mental turbulence of anxious, avoidant, and disordered attachment is addressed. As evidenced in my recovery-group experience, this can manifest later in life in anxious and co-dependent relationships or an obsessive preoccupation with relationship itself. At the physical level, relationship manifests in the activities of the endorphins, phenethylamine, oxytocin, serotonin, and dopamine.

What you can do now: When we are insecurely attached in the early stages of life, we often perceive ourselves as being abandoned. Because we are helpless at that age, we are dependent on caregiver relationships for our very survival. Later in life, those who were insecurely attached can see the establishment of relationships as a matter of life and death. In such individuals, the fear associated with abandonment is encoded somatically (in their body) and manifests from deep within the unconscious. If you are such a person, now is the time to engage your compassionate curiosity in *making the unconscious conscious* in this respect. Are you only content when pursuing or engaging in novel relational behaviors? Keep an open mind and consider every possibility. Now is also the time to work on building a secure attachment with your higher power, the one with the key to our ultimate survival.

7

Interoception

"The feeling of knowing what is happening in your body, for example if you are hungry, thirsty, warm, cold, etc." —CAMBRIDGE DICTIONARY

THE ONLY TOOL WE need when exploring internal phenomena is *interoception*. This is the process of consciously observing the sensations and emotions within our body, it is what I have referred to so far as *self-awareness*. Interoception is easier to use for some than others, but I encourage everyone to learn to pay attention to what's going on inside in a spirit of non-judgment. The behavioral challenges we face are mostly unconscious responses to internal feelings that we might not even be aware of. The limbic loop is an expert at turning those feelings into actions, both good and bad.

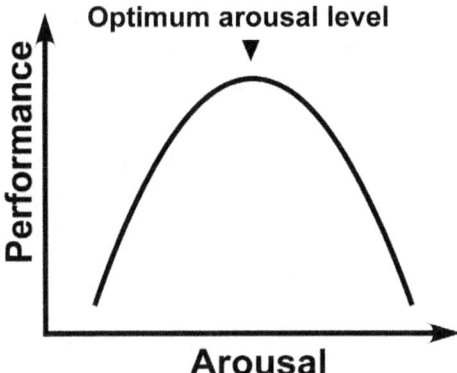

Figure 7.1 The inverted-U. Increased Arousal Correlates with Increased Dopamine and Norepinephrine. Wikimedia Commons.

Common sense suggests that a higher level of dopamine signaling in the limbic loop would result in a more intense reward. This is true to a point, but when we graph the relationship between dopamine levels and the reward experienced, we see an inverted-U shaped curve rather than one that continues to rise indefinitely. The reward experienced increases with the upswing of the curve until it reaches its peak, after which feelings of well-being are progressively replaced with those of overwhelm and anxiety. Several laboratory studies illustrate this principle using dopamine boosting drugs. Most of these studies discuss the curve's effect on cognitive performance through the action of dopamine on receptors in the pre-frontal cortex, but the same effect manifests in the limbic system. I can speak to this from a phenomenological perspective having been prescribed both the dopamine re-uptake inhibitor Ritalin and the dopamine inducer dextroamphetamine as treatments for adult ADD. If my doctor tells me to double the dose under certain conditions, I can sense the change from rewarding feelings early in the day to more anxious ones in the afternoon. Anything after the second dose in the morning does nothing to enhance my mood even though it does help me maintain focus. Understanding these felt-sense experiences (feelings) and the way they influence our behavior helps us to let go of the blame we ascribe to ourselves when we fail to live in accordance with our values. It provides a *why* to *what* of our actions. We experience some of our most dysphoric states when our esteem needs are in arrears, and this is when we are most likely to act out.

The Role Feelings Play in Our Lives

In his book *Feelings: The Need for a New Science*, Gyozo Margoczi[1] addresses the lack of emphasis placed by science on feelings and needs, even though these are fundamental underpinnings to the human experience. The birds-eye-view perspective of someone from another specialty is often required to tackle subjects that are not in vogue in their respective fields of study. Margoczi approaches the subject of feelings from an engineer's perspective and uses mechanical analogies to illustrate his points. Appetite, the impetus that drives us to meet our needs, is a complex process involving many regions of the body and brain. These regions differ depending on the need in question, but the following illustration from Margoczi's book gives a unified view of appetite that he believes can be applied in all cases.

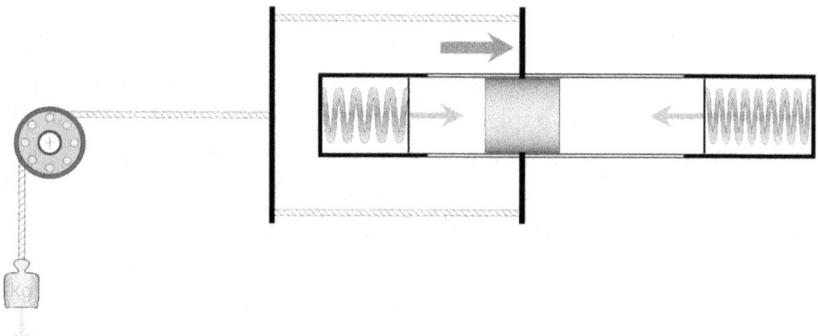

Figure 7.2 The Inverted-U, a Mechanical Analogy.
Gyozo Margoczi, used with permission.

On the right of the illustration in Figure 7.2 is a tube with springs at both ends. The tube contains a piston that is connected to ropes via two splines which run in slots cut in the top and bottom of the tube. On the left a rope connects the splines on one side and a weight on the other. The rope runs over a pulley and over time the weight draws the piston to the left by the force of gravity. Eventually the piston meets the leftmost spring. As gravity moves the piston further to the left, the spring becomes compressed. This tension indicates the dysphoria that accompanies unmet need. As the intensity of this sensation increases the call to action grows, and the individual feels compelled to perform some task (ingesting

1. Margoczi, *Feelings*, 48.

food, drinking water, breathing air etc.) The meeting of this need drives the piston to the right and into the comfort zone. However, if it is moved too far to the right (for example by eating too much food, drinking too much water, or hyperventilating) it meets the spring on the right. This spring then becomes progressively compressed as the individual ingests more of the materials of life, resulting in another type of dysphoria. If this continues unabated, death will result due to a ruptured gut, Hyponatremia or Respiratory Alkalosis, respectively.

If the goal of a behavior is the direct production of rewarding feelings, however, the individual might perceive the decrease in those feelings beyond the peak of the inverted-U as a reason to continue the behavior instead of curtailing it. This is done in the hope that those feelings can be recaptured before the end of the session.

The Inverted-U Model of Dopamine Function

While the negative physical aspects of overconsumption are multifactorial, the incentive to consume more normally ceases when tension on the right-hand spring starts to build. Dopaminergic function is not the only contributor to this state of being, but it plays a key role. The link between this principle and the inverted-U of dopamine response is an area ripe for further research as it is central to an understanding of addiction. As with everything dopamine related, however, the situation is a complex one. Certain drugs that raise dopamine levels, such as phentermine and the amphetamines, are used as appetite suppressants—they promote anorexia. This may be because the levels of dopamine have been pushed past the peak of the curve and have taken the incentive to eat along with them. It may also be because of the adrenergic effects of these drugs, as adrenaline (epinephrine) is an appetite suppressant. In contrast to this, the cannabinoids raise dopamine levels but often *increase* appetite (a condition known as *the munchies.*)

The peak of this curve may also be accompanied by an increase in cognitive function (due to raised dopamine activity in the pre-frontal cortex) giving certain drugs, such as Modafinil, the epithet "Smart Drugs." The effect of Alcohol on cognitive performance may also illustrate this principle.[2] During the year of practical experience for my undergraduate degree, I worked for the Human-Computer Interface Research Unit at

---

2. Castillo, "Can Alcohol Make."

Loughborough University. One of the team members shared with me his strategy of using alcohol to improve his programming performance. I was skeptical that this was possible, but the research seems to support the finding. The principle was given a satirical treatment in the webcomic xkcd using the concept of the Ballmer Peak—named after Steve Ballmer, the president of Microsoft from 2000–2014. Figure 7.3 illustrates their extreme take on the inverted-U response curve.

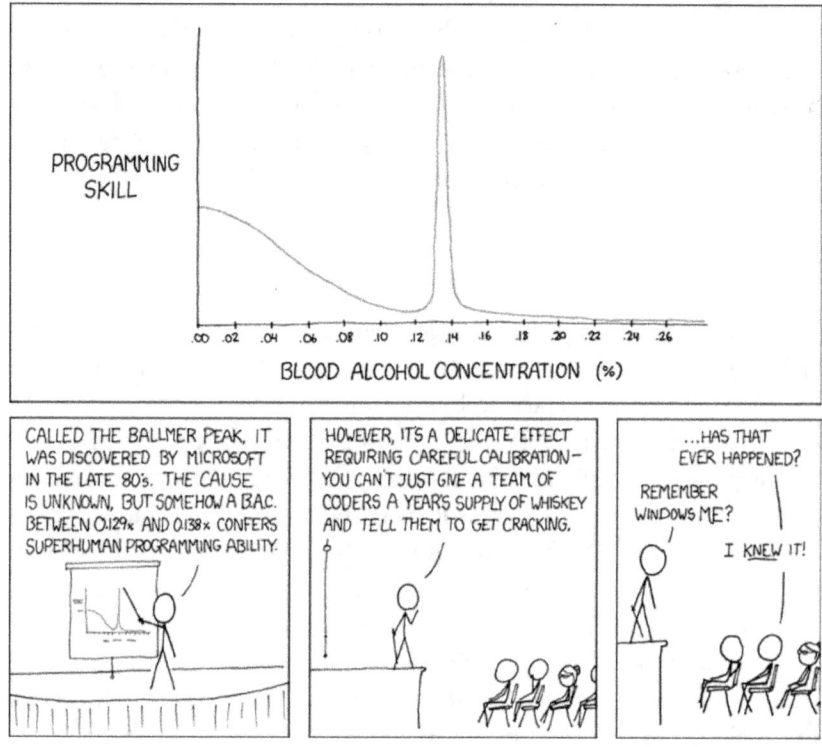

Figure 7.3 The Ballmer Peak. xkcd.com, used with permission.

Alcohol is a tiny molecule with a complex profile that affects the GABA, glutamate, and opioid signaling pathways, among others. Research has shown that its primary effect on the limbic reward pathway is through the release of endorphins. Because of this, its limbic effects can be blocked using the mu opioid receptor antagonist Naltrexone.

Double (Or Multiple) Doping

The decline in positive effects following the sweet spot in the inverted-U works against the goals of the addict. If the levels of the drug of choice in their system have risen past the point of maximum reward, further ingestion is pointless, but the allure of the high remains. This tends to promote experimentation with additional drugs that counteract the effect of the primary and return the user to the peak of the curve. In his memoir *On Writing*[3], Stephen King discusses his addiction to cocaine and the lengths he would go to in chasing the high, which included taking Benzodiazepines, drinking NyQuil, and chugging mouth wash. His cocaine habit eventually had such a negative effect on cognitive function that he can hardly remember writing the best parts of his novel *Cujo*. King reformed after an intervention by his family, but another "King" whose drug use contributed to an early death was not as fortunate. His story illustrates how far double doping can escalate.

Elvis Presley's final autopsy report identified over a dozen stimulants, pain killers and sedatives in his system at the time of death. The concentration of Codeine (methyl morphine) alone was ten times the therapeutic dose. In her autobiography *Elvis and Me*[4], his wife Pricilla Presley describes his rampant drug use. By the time she met him it had already advanced beyond the quest for reward into a deeply ingrained dependency on substances for normal function. His tendency was towards living in dark rooms during the day and being dependent on drugs for both sleep and wakefulness. It is easy to judge Elvis for his lifestyle, but in an article entitled *Elvis Presley: Head Trauma, Autoimmunity, Pain, and Early Death*[5], Medical Doctor Forest Tennant puts the blame for Elvis's addiction on a string of concussive events rather than his character deficits. This type of diagnosis has gained acceptance in mainstream medicine since the identification of Chronic Traumatic Encephalopathy (CTE) amongst professional athletes. It is believed that CTE results from the cumulative effects of multiple concussions received through accidents or while engaging in contact sports. Its symptoms include confusion, aggression, and a lack of impulse control which often leads to drug abuse. Death is often the result of suicide. Ergo, an individual's drug abuse can result from circumstances that are quite beyond their control.

3. King, *On Writing*, 97
4. Presley, *Elvis and Me*, 212.
5. Tennant, "Elvis Presley."

The Inverted-U In Daily Experience

The inverted-U also shows itself in aspects of our everyday experience. In chapter 8 I discuss the effect of money and material goods on dopamine activity, and it is here that the deceptive nature of limbic reward shows itself clearly. Material goods meet needs and as a result promote feelings of wellbeing. You would think that feelings of happiness would increase with the accumulation of possessions, but when we graph the amount of happiness experienced by people in various socio-economic groups, we find there is no absolute correlation. The peak in happiness for those who use money as an emotional support occurred at around $100,000 per breadwinner.[6] After that, the graph flattens out, and in some cases happiness and increased income were negatively correlated. For example, there are cases where the thrill of winning the lottery was short-lived. One recipient complained that his stress level had *increased* since the win due to anxiety over the performance of his investment portfolio. As Brené Brown writes in *Daring Greatly*[7], "The opposite of scarcity is not more than you can possibly imagine, the opposite of scarcity is *enough*." I might paraphrase this as "You can't get a bigger reward than the biggest reward your brain can give you, and that reward is not as expensive as you might imagine."

Sadly, the human psyche is not familiar with this concept and in some cases will continue to strive for more despite having exceeded the peak of the dopamine response curve. Judith Grisel, in her book *Never Enough: The Neuroscience and Experience of Addiction*[8] takes this to the next level. Prior to becoming a neuroscientist Grisel was a cocaine addict and in late 1985 a large amount of the drug fell into her possession by accident. Despite having more of the drug than they wanted, Grisel and a friend proceeded to use all of it. This would have taken them far beyond the peak of the inverted-U if that were even a possibility, but due to the neural adaptations they had both accrued following years of cocaine abuse, their dopamine response curves were essentially flat. They were beyond using cocaine as a driver of reward and saw it merely as a means of escaping reality. After the larger-than-expected bag was empty, her friend announced, "there would never be enough cocaine for us." This perspective is mirrored in the billionaire who has more

---

6. Berger, "Does More Money."
7. Brown, *Daring Greatly*, 29.
8. Grisel, *Never Enough*, 1.

money than they could spend in many lifetimes but is still driven to accumulate more wealth. Karl Marx ascribed Biblical importance to this drive in the lives of the rich:

> Accumulate, accumulate! That is Moses and the prophets! . . . Accumulation for accumulation's sake, production for production's sake.[9]

Novelty

The relationship between reward and *novelty* also plays a role in the addiction process. As we have seen, over time, exposure to any drug (and some addictive behaviors) is countered by physiological adaptations in the CNS. These are the body's attempts to regain equilibrium in a system that is now out of balance. These adaptations might include an increase or decrease in the expression of receptor types, the deactivation of existing receptors, an increase in the expression of enzymes that neutralize messaging molecules or the production of endogenous antagonists (for example, anti-opioids,) among other approaches. Because these adaptations reduce a drug's effectiveness, the inverted-U is moved to the right with respect to the concentration of the drug. As it moves to the right, the height of the peak is also reduced. Eventually, high concentrations of the drug are required to maintain normal function, and the height of the peak is minimized. The dysphoria to the left of the peak is also intensified, which increases the incentive to ingest the drug or engage in the rewarding behavior.

The difference between the worn-out recording of a favorite song and an adapted neural network must be noted. Losing interest in a song that was once our go to comforter in times of trouble is not the result of receptor adaptation. If that were the case, we would lose interest in all music after a prolonged period. However, novelty is a key mover in both cases. Novelty decreases in relation to time and repetition. We certainly see the novelty principle at work, not just with music, but also in our experience with foods and people. Our favorites in each of these categories can lose their allure over time. While a sense of well-being for some people can be maintained with little input from the outside, others continually seek novel activities to remain in their comfort zone. This adaptation to the familiar is also key to the addiction process.

9. Marx, "Capital Vol. I - Chapter Twenty-Four."

A Closer Look at Drug Adaptation

In her book Judy Grisel also presents a distinct perspective on drug adaptation using the A process/B process model. The A process represents the increase in drug-mediated function relative to the normal baseline of everyday life. The B process is the brain's opponent response to the drug which attempts to return neural function to normal levels.

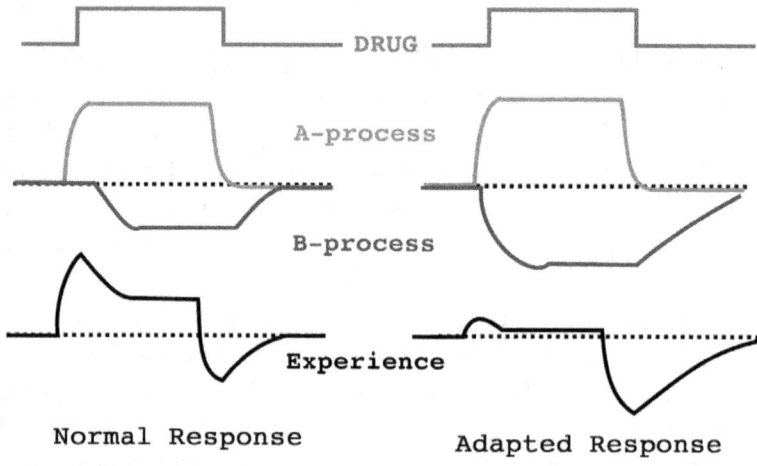

Figure 7.4 The Opponent Response to Drugs. Wikimedia Commons.

The experience line on the left of the diagram above represents a new user's subjective state in response to ingesting the drug. The initial response is a euphoric high, which levels out for a period and then dips below the baseline, producing a dysphoric low. The low resolves over time as neurological function returns to normal. The low is inevitable but is factored into the user's drug usage plan as part of a cost-benefit analysis. Those who have empathy for their future selves will moderate their use in anticipation of this low, which might help ameliorate the negative effects of drug use overall. Those whose only consideration is the present moment will enter the adapted response phase, illustrated on the right, much more rapidly.

After adaptation, the A-process on the right is slightly elevated because the user has ingested more of the drug. However, the experience line, indicating the subject's mood, now barely rises above the baseline. Due to the adaptations that have taken place in response to repeated administration, the B-process takes effect more rapidly and aggressively than before and almost completely counteracts the subjective effects of the A-process. Despite the absence of any euphoria at the start of the cycle, the dysphoric low is even more intense, a fact that encourages the user to administer the substance repeatedly throughout the day.

While this example involves an exogenous drug (one introduced from outside the body,) a similar effect can be observed with endogenous drugs (those produced inside the body.) In chapter 1 I examined the link between the body's own amphetamine-like compounds and the relationship phase known as infatuation. In this state, the rewarding sensations resulting from the actions of phenethylamine and oxytocin on the nucleus accumbens cannot be maintained indefinitely, as the body always returns to a state of equilibrium. The decline of the infatuation experience, however, stems from a reduction in situational novelty rather than physical adaptations at the receptor level. Physical adaptations to endogenous drugs are hard to produce without prolonged engagement in powerfully rewarding behaviors where novelty can be maintained indefinitely, such as viewing Internet porn.

The Zuckerman Sensation Seeking Scale

Susceptibility to low-dopamine states is not equal across the entire population, with some finding easy access to contentment while others are driven to engage in novel behaviors that might endanger both their physical health and their relationships. In the 1960's Marvin Zuckerman developed a four-axis scale (the Zuckerman Sensation Seeking Scale[10]) to classify personality traits such as neuroticism, antisocial behavior, and psychopathy. The four axes are:

Thrill and Adventure Seeking (TAS)

Disinhibition (DIS)

Experience Seeking (ES)

Boredom Susceptibility (BS)

Sensation-seeking and novelty-seeking are highly correlated. Zuckerman proposed that these traits originate in some unknown psycho-biological system. We can intuitively see that each of these traits

---

10. Zuckerman et al., "Sensation Seeking Scale."

involve a desire to escape some sub-par state of being by introducing a form of external stimulation. This is consistent with a state in which a subject's limbic activation falls within the left-most quartile of the inverted-U. States of being, which to the average individual fall within the comfort zone of the curve, are for these people at least mildly dysphoric. Boredom is a feature of this state that we can all relate to. Most parents are familiar with the childhood lamentation "what can I do"? uttered, as it usually is, in a despondent tone. When we suggest a solution (some form of housework like "tidy your room") the child rejects the suggestion, knowing that this form of activity is unlikely to budge the dopamine curve from its second-rate position. They are, however, incapable of formulating a course of action that will do the trick. In the absence of an acceptable solution, however, there is always one approach a parent can use as a wild card.

Punishment demonstrates that some dopamine-inducing activities are socially unacceptable; it also breaks the cycle of their expression. Punishment replaces the sought-after reward with some form of pain or deprivation. But the canny individual knows that their transgressions will go unnoticed, given a little luck. This introduces a dopamine-inducing "maybe" into the equation. The individual will assess the thrill of the activity and the dopamine hit of evading detection against their prior experiences with pain and deprivation to determine whether the benefits outweigh the risks.

This may also explain the behavior of people with attention deficit hyperactivity disorder. Because their dopamine response curve is normally suppressed, much of their mental activity is consumed in dopamine-inducing activities and thoughts. This line of reasoning would also explain why drugs that raise dopamine levels, such as Ritalin, dextroamphetamine and methamphetamine are effective in the treatment of ADD. My conversations with fellow Ritalin users identified a state of calm in which the individual is content to remain focused and inactive, not through the sedation one might experience from a tranquilizer, but because of appetite satiation.

Sensation Seeking and the Inverted-U

Sensation-seeking is something that declines with age in both men and women, as does the density of dopamine receptors in the limbic regions of the brain. This observation (related to a reduction in sex hormone levels) prompted research into the relationship between the Zuckerman Sensation Seeking scale and dopamine receptor density. Prior research also shows that sensation-seeking behavior is inversely correlated with dopamine receptor availability (as opposed to dopamine receptor density.) At any point in time, some dopamine receptors will be bound to a dopamine molecule, and others will be available for binding should dopamine be released into the Synaptic Cleft. In the unlikely event that all receptors are bound, the further release of dopamine will have no effect on neural signaling. This is the effect experienced in the post-apex downward sweep of the inverted-U.

In a 2010 paper titled *Inverted-U-shaped correlation between dopamine receptor availability in striatum and sensation seeking*[11], a team of researchers used Positron Emission Tomography and the radioactive dopamine receptor antagonist Carbon-11-Raclopride to map the brains of eighteen male subjects. The subjects had previously been categorized based on their level of sensation-seeking behavior using the Zuckerman scale. The results indicated a statistically significant correlation between dopamine receptor occupancy and sensation seeking. Higher dopamine receptor occupancy increased sensation seeking. This research also explored the concept of "Gain." In the audio recording world, the term refers to the volume level of a particular audio input or output. In the case of neurotransmission, it describes the level of electrical activity induced in the downstream neuron by the release of a particular neurotransmitter. Each dopamine receptor, in the receiving neuron, that moves from the unbound to the bound state will contribute to the gain generated by the flow of positively-charged ions across the neuron's cell membrane. However, if the receptor is already bound, it is unable to contribute to this gain. Sensation-seeking is, therefore, not related to low dopamine levels in the synapse, but rather the combination of dopamine levels and the number of available receptors. A high base-level binding count effectively moves the U-shaped response curve to the right and puts contentment and satisfaction beyond reach unless the subject engages in some novel experience. The sensation seeker is in a similar state to the drug addict whose brain has already adapted to higher dopamine levels and has become less responsive (due to a reduced dopamine receptor count.)

So, what does this mean for the sensation seeker? Every day they are likely to feel uneasy, edgy, out-of-sorts, restless, and unsettled (all

---

11. Gjedde et al., "Inverted-U."

symptoms of dopamine withdrawal.) They may tend towards depression and see life as meaningless and unfulfilling unless they engage in dopamine-boosting activities that they might prefer to avoid. This has implications for the believer in their relationship with God. God provided a moral code that prioritizes the needs of others over those of the self, but an individual's ability to contribute to others relates directly to their own mental health. The sensation seeking brain is needy—it requires a high energy input to function at normal levels, and this leaves less energy for the business of living and giving.

The Developing Limbic System

Due to the elementary state of pre-frontal cortex development in the brains of children, they are more susceptible to the limbic influence than adults are. There is, however, one field where limbic plasticity expresses itself more forcefully in adults as we shall see. For a child everything is new and exciting (novel); they have powerful drives for play, attention, nourishment, and a burning self-interest. These drives express the wisdom of the developing brain—without them the child would not survive or learn. A child's limbic processes operate in virtual isolation from the higher functions of the pre-frontal cortex. For the first eighteen years of life, the role of pre-frontal cortex falls on the parent. Their job is to instill in the child's memory the behaviors and cultural norms required to function in the societal context. When a person is sufficiently mature, they become capable of self-regulation and no longer need parental support in the basic things of life.

The drive towards this state of independence, however, begins long before maturity sets in. The trigger for this drive occurs in the pre-teen years and its influence is quickly expressed in the teen obsession with members of the opposite sex. Children possess the same amount of genetic material at birth as they do in adulthood, but certain genes remain dormant until they are triggered by the release of sex hormones. We are familiar with the basic physical changes that occur at this time, but there are also unseen adaptations taking place in the brain as the result of Testosterone and Estrogen production. One of these involves the proliferation of dopamine receptors in the limbic reward pathway.[12] This is responsible, in part, for the profound changes in response to visual and tactile stimulation

---

12. Purves-Tyson et al., "Testosterone Induces."

that we describe as *attraction*. This development also rewrites the rules of motivation and behavior in many areas related indirectly to reproduction such as emotion, autonomy, and aggression.

The experimental use of substances also begins for many at around this time. In my day this was largely limited to nicotine and alcohol use. These rites of passage produce a false self-perception of maturity in those who partake and a belief that those in their social orbit will share this perception and find them attractive or worthy of recognition (esteem needs.) This perception alone could render these compounds habit-forming, but the changes in the state of consciousness they promote add powerful impact to the experience.

One line of research illustrating the link between androgen (sex hormone) levels and dopamine receptor density is the study of psychosis. Most anti-psychotic drugs are dopamine antagonists, and psychosis is very rarely diagnosed until several years after puberty. Antipsychotics are effective, so the theory goes, because psychosis is caused by overactive dopamine signaling.[13] This overactive state cannot be reached until dopamine receptor density has achieved the necessary threshold in response to long-term androgen receptor activation.

Non-Violent Communication

A boost in self-regard will directly meet esteem needs, but it is through dopamine that this change is communicated to the consciousness using the language of feelings. In his book *Non-Violent Communication: A Language of Life*[14], Marshall Rosenberg explores this language. He proposes that if needs met, positive feelings result and if they remain unmet the emotional fallout is negative. Feelings are the impetus that drives need-satisfying behavior. Overlaid on the NVC model is the concept of *empathy*, the ability to enter the experience of another, take their perspective, walk a mile in their shoes. Without empathy the feelings aroused by unmet needs, in a relationship, trigger the use of aggressive or passive-aggressive behaviors and language. Rosenberg describes this as "jackal language" speaking analogically of a small canine inhabitant of Eurasia and Africa that exhibits typical wild canine behavior. As we will see, this jackal spirit expresses itself in the human through the *amygdala*,

---

13. Howes and Kapur, "The Dopamine Hypothesis."
14. Rosenberg, *Non-Violent Communication*.

also known colloquially as *the guard dog*. Using NVC this jackal can be subdued through an understanding that both parties have the same basic needs and that a solution can be found in which the needs of both are met. The most life-serving solutions involve cooperation rather than competition. They involve power-with rather than power-over.

NVC is not just an idea, but rather a state of consciousness that Rosenberg championed world-wide until his passing in 2015. He mediated conflicts between Israelis and Palestinians, and between warring tribes in Africa. He established schools all over the world that operate based on heart-felt connection between individuals rather than the suspicion, judgment, and conflict commonly seen in those institutions. He described NVC as a language of life and, in so doing, emphasized the expressive value of language in communicating internal states of being between individuals. He created wordlists for both feelings and needs that reveal the power of language in this process, and stressed how much language has evolved precisely for the expression of these ideas. During the initial development of NVC he labelled needs as *wants* but as discussed in the previous chapter, an expression of want does not have to be a sign of toxic neediness and negativity. A want could reveal a valid impediment to emotional and functional balance within the individual. Most needs are things we can live without, but which are necessary for us to feel "ok." Here are some examples of feelings we experience when needs are met, and those we encounter when they go unmet.

Needs Met (Stimulation)	*Needs Met (Sedation)*	*Needs Unmet*
Energetic	Calm	Apprehensive
Enthusiastic	Comfortable	Frightened
Invigorated	Centered	Bewildered
Vibrant	Content	Perplexed
Eager	Fulfilled	Torn
Fascinated	Mellow	Alienated
Entranced	Satisfied	Detached
Empowered	Serene	Uninterested
Jubilant	Tranquil	Withdrawn
Delighted	Relieved	Disgusted

Needs Met (Stimulation)	Needs Met (Sedation)	Needs Unmet
Blissful	Equanimeous	Contempt
Ecstatic	Still	Hostile
Elated	Open-Hearted	Outraged
Thrilled	Tender	Irritated
Rapturous	Warm	Exasperated

Table 7.1 Feelings arising when needs are met/unmet

The words in the center and left columns illustrate the operation of our reward circuitry in response to needs being met. Overall, our minds and bodies know what they need to operate at full capacity, and many human needs relate to the improvement of our life circumstances. In Table 7.2 I have listed some of the needs identified in the NVC needs inventory which, when met, bring about a reward response.

Basic Human Needs

Acceptance	Belonging	Consideration
Inclusion	Authenticity	Integrity
Equality	Beauty	Order
Freedom	Independence	Justice
Competence	Creativity	Learning
Understanding	Clarity	Spontaneity

Table 7.2 Some needs that must be satisfied to promote optimum mental health

We can use our craving for air to illustrate the traditional definition of a need (something that is required to keep us alive.) None of the needs cited above fall into that category. We will not die if we lose our independence by being held in solitary confinement, neither will we die if our creativity is thwarted by a monotonous and mundane line of employment. However, these things will affect our quality of life.

This idea is further explored by the Enneagram—a system of personality type classification that assigns each individual one of nine personality types and positions them on a nine-level mental health scale.

This scale represents nine subdivisions in the gamut of subjective experience, spanning the space between the high-performing individual and the catatonic one. The latter is often identified in colloquial English as a state of "Nervous Breakdown." Viewed in this way, meeting needs is not so much a matter of life and death, as it is a pre-requisite for healthy mental function. As the list of an individual's unmet needs grows, so the feelings associated with their lack progressively show themselves, as do the behaviors associated with those feelings. Perplexity, irritation, and exasperation do not make for peaceful communication.

The Role of Unmet Needs in Promoting Addictive Behaviors

Table 7.2 identifies eighteen unique needs—a subset of those mentioned in Baynvc's Needs Inventory. Even if we restrict our consideration to these eighteen, it is unlikely that all of them will be met at any point in time. With some consideration of the emotions sensed within us, and observed in others, we will notice that we experience many feelings from the leftmost two columns of table 7.1 (at least temporarily) when only one or two needs are met. In rewarding us with positive feelings for enhancing our life circumstances, our mind is promoting a healthy lifestyle. If needs remain unmet, our mind will use feelings like those in the rightmost column of Table 7.1 to indicate that something needs to change.

If our needs remain unmet for extended periods of time, we may conclude that the situation is hopeless and might attempt to recreate the *feeling* of met needs using substances or other non-life-serving behaviors, such as our involvement in multiple out-of-integrity intimate relationships. This approach is not actually meeting needs but erecting an artificial barrier between ourselves and the negative feelings we are trying to escape. Strategies like these draw the individual away from healthy function and towards breakdown.

It is quite clear that the first two columns of Table 7.1 describe different aspects of dopamine flow and limbic reward. I have categorized the positive feelings into stimulation or sedation to show the similarity between these expressions of emotional state and the subjective effects of stimulant and depressant drugs. The left most column heading could read Ritalin/meth amphetamine/ecstasy. The middle column might be labelled Oxycontin/Ativan/Ketamine. I do not mean to minimize or objectify the feelings themselves, but to show that the sensation of met needs is mediated by the same neural infrastructure targeted by drugs of abuse. Substance abuse is an attempt to emulate those interoceptive states which are normal in healthy individuals, but are out of reach for some, due to their biological makeup, life circumstances or prior traumas.

Thematic Takeaways

Feelings: The feelings experienced when needs are met mirror those produced by stimulant and sedative drugs. Our addictive attachment to chemical and behavioral resources is a strategy to create the states that arise naturally when needs are met. We seek these artificial solutions when needs remain unmet for extended periods of time.

Needs: While dopamine drives us to meet a vast array of needs and urges us to endlessly crave more of the resources that meet those needs, this craving is deceptive in nature. The inverted-U principle reveals that the most we can ever gain is enough, and that further needs-satiating behaviors provide diminishing returns and often the loss of the reward we seek. The phrase "less is more" encourages us to question whether more is better. When speaking of dopamine, more is only better until the peak of the inverted-U has been reached.

Neural Activities: The positive effects of psychostimulant drugs are not directly related to the amount consumed. In the short term, an increase in cognitive performance and mood may be experienced to a point after which the benefits of further ingestion will decline, making further consumption pointless. Following long-term exposure, the CNS adapts to exogenous drugs and some behaviors to protect against neuro-toxic levels of electrical activity.

What you can do now: We generally crave dopamine/nucleus accumbens stimulation under two conditions, when dopamine levels are low (when needs are not met) and when they are high. When dopamine levels are high we enter a celebratory state and want to engage in activities that release more dopamine. Healthy acts of celebration contribute positively to overall health, but if we have disavowed a particular behavior, perhaps smoking cigarettes, a celebratory atmosphere is likely to trigger a nicotine craving. Start observing this phenomenon in yourself. When you are having a good day and feel drawn towards an old vice, it is just dopamine doing its thing. The craving is not an aspect of your true self but evidence of the destructive strategy to which you have become fused. Practice perceiving the craving as separate from your identity and see the benefits.

8

Money

"Money has never made man happy, nor will it, there is nothing in its nature to produce happiness. The more of it one has the more one wants." —BENJAMIN FRANKLIN

ALL MAMMALS HAVE A limbic system and experience a similar range of emotions to humans. They are also rewarded by dopamine activity, and like us it is often from the anticipation of an enjoyable experience. My dog is less than twelve inches (thirty centimeters) tall, but when he sees me pick up his leash, he jumps so high he can nip my jacket between my shoulder blades! He becomes highly stimulated when anticipating the reward of sniffing the "messages" other dogs have left and leaving his own calling cards.

A study on monkeys, described by Robert Sapolsky[1], demonstrated a similar reaction to that shown by my overexcited pooch. While the primates tested were not humans, humans are primates and the similarities between human brains and primate brains indicate that the same processes are at work in both species. The experiment established a clear correlation between dopamine levels and the incentive to act.

The monkeys were trained to respond to a light which indicated the start of a feeding session. Following its illumination the animals pushed a lever ten times (the labor required to receive the reward,) after which food was dispensed. The dopamine levels in the monkey's brains were measured at each point in this process, and the levels spiked as soon as

1. Sapolsky, "Dopamine Jackpot."

the light came on and remained high until the food was dispensed. The high dopamine levels indicated a state of anticipation. If a dopamine antagonist was administered prior to the test, the behavior was completely absent. The dopamine spike was the incentive to perform the task and without the incentive there was no reason to act.

The Role of Dopamine in Movement

Dopamine plays a role in anticipation and action, capacities we all need to perform our daily labor. I learned about dopamine's involvement in movement first-hand shortly after taking a single Seroquel (Quetiapine) pill to help me sleep (something I never did again.) Seroquel is a dopamine antagonist that conflicts with my sensitive system. I awoke with a bout of nausea that built to the point where a trip to the bathroom seemed in order. I made it to the door, but when I told my arm to operate the handle, nothing happened. This was followed by a loud noise which I didn't realize at the time was my body hitting the hardwood floor. My brain was unable to form the intention to raise my arm because dopamine signaling in the Substantia Nigra (the brain region where the intention to move originates) had been blocked. I was also unable to maintain muscle tone in my legs which resulted in their collapse. Seroquel is known to have this effect, but usually only after years of high dose use.

Movement is a central function of the CNS. One model of CNS activity assigns movement such a pivotal role that language itself is seen as a sequence of actions, a coordinated progression of muscle contractions in the mouth, throat, and larynx. When viewed in this way, even the self-talk in our heads becomes a pattern of actions, although these are merely planned and not executed.[2] This model views the brain as having many inputs but only one output—movement.

The Santa Clara "Frozen Addicts" Case

The role of dopamine in movement was brought into stark relief in a 1982 case[3] involving six heroin users. Over the course of several days each of these patients were brought into the Santa Clara Valley Medical Center with the symptoms of severe Parkinson's disease. None of them were able

---

2. House, *Nineteen Ways*, 109.
3. Locklear, "How Tainted Drugs."

to move or even talk. Curiously, they were all under forty-five years of age (one was only twenty-one,) and Parkinson's usually develops after the age of sixty. Parkinson's disease is the result of cell death in the substantia nigra. These cells use dopamine as a signaling molecule, making it how the electrical signals involved in movement are initiated.[4]

Medical staff at the hospital were baffled by this tragic and unusual situation but eventually identified a cause. Each of the patients had intravenously administered a synthetic opioid called MPPP for its rewarding effects. However, the chemist that manufactured the MPPP had made an error, resulting in a small amount of MPTP being included in the finished product. MPTP has a structure like that of dopamine so it can be absorbed into the pre-synaptic neuron by the dopamine transporter. Once inside the cell, MPTP is broken down by mono amine oxidase B (MAO-B) and the sub-components (metabolites) of this deconstruction process can have deleterious effects on the neuron. The metabolite that caused the problem was MPP+. This proved to be toxic to the dopamine neurons, causing many of them to die off.

Attempting to return these patients to normal function, doctors treated them with levodopa (the raw material from which dopamine is synthesized within the body.) The goal was to return Substantia Nigra function to normal levels, but instead their bodies were seized by the uncontrollable movements known as Tardive Dyskinesia.[5] This condition is experienced by Parkinson's Disease patients, acutely, when treated with high doses of levodopa, or cumulatively with the low doses. Two other drugs used in Parkinson's treatment are Pramipexole and Ropinirole, both dopamine agonists. Remarkably when some people commence treatment with these drugs, they become addicted to gambling.[6] The increase in dopamine signaling stimulated by these drugs is designed to improve the functioning of the Substantia Nigra, but they also amplify dopamine activity in the incentive and reward pathway. The extra activity in the limbic loop stimulates such a craving for the win that the incentive to engage in risky financial behavior becomes irresistible.

Nicotine also activates the limbic system and this is the principle that drives cigarette addiction. Nicotine binds with acetylcholine receptors in the ventral tegmental area resulting in an efflux of dopamine

---

4. Aans.org, "Parkinson's Disease."
5. Pandey and Srivanitchapoom, "Levodopa-Induced Dyskinesia."
6. Scott, "Parkinson's Drugs."

in the nucleus accumbens.[7, 8] Again, the result is a rewarding state. I witnessed a curious vignette during a lunchtime walk a few years ago that caused me to wonder about the effects of combining levodopa and nicotine. As I walked along a cycle path in my hometown, a man in his sixties or seventies approached me on a motorized wheelchair. It was impossible to ignore the exaggerated movements he was making throughout his body. Despite this he was able to steer the wheelchair perfectly well with his left hand while he held a lit cigarette in his flailing right hand. I assumed the dyskinesia was a side effect of levodopa treatment and wondered why he would be smoking while in that state. My guess is that there were synergistic effects between the two drugs that produced a powerfully rewarding response. This double doping strategy is common among drug users as is the abuse of prescription drugs among the general public.

Money as a Storage Container for Dopamine

Another monkey study, conducted by Frans de Waal[9] explored the relationship between performance and reward. This involved placing two Capuchin Monkeys side by side in cages in clear view of one another. The monkeys would happily perform the task of selecting a rock and handing it to the researcher in return for food. The selection of the rock was the *labor* that the monkey performed to earn its reward. If both monkeys were given cucumber as payment for their labor they would continue to perform the task repeatedly, however, if one was given cucumber, and the other a grape, the subject with the less valuable reward became indignant. In one case, after receiving the lesser reward the first monkey threw it at the researcher and shook the bars of the cage violently, a response that occurred twice in a row. After showing the video at a conference, de Wall commented "So here we see the Occupy Wall Street protests." This behavior indicates that the need for fairness, equality, and justice exist even in the animal kingdom.[10] Professor Lixing Sun covers these needs and their neurological underpinnings in his book *The Fairness Instinct: The Robin Hood Mentality and Our Biological*

---

7. Wittenberg et al., "Nicotinic Acetylcholine Receptors."
8. Benowitz, "Pharmacology of Nicotine."
9. De Waal, "Two Monkeys."
10. De Waal, "What is an."

Nature.[11] The Bible also illustrates this reaction in Matthew 20:1–16 (The Parable of the Workers in the Vineyard.)

In the context of their daily labor, however, humans are generally rewarded with money rather than food and this brings us to a key point. We have already seen that the monkey's behavior is suppressed by a dopamine antagonist. Behavior is therefore a result of the incentive to act and the ability to act, both of which are mediated by dopamine signaling. When dopamine provides the incentive to perform a task it is anchored in the corresponding reward, but humans have an abstract representation of reward, one that has almost infinite flexibility.

When the monkey was rewarded with a grape all it got for its labor was physical sustenance and a tasty treat. When humans are rewarded with cold hard cash, they are given agency to select the reward they desire. It might be food, clothing, healthcare products, toys or any other commodity or service. Money is both a symbolic representation of reward and a means of storing reward over time. It is similar in nature to potential energy. When we ride our bike to the top of a hill it is hard work, but we get to redeem that work on the way down the other side. The potential energy stored in the mass of the bike and the rider's body is a function of altitude and gravity. With money our work may first be vested in a promise or contract with our employer for the work rendered. When pay day comes, we receive a symbolic representation of that work which we use to meet our physical, safety, security, and esteem needs, among others. The more money we earn the more ease we have in meeting these needs. We exist in an incentive/reward cycle:

1. Dopamine moves us to engage in creative activity with the anticipation of a reward.
2. We receive remuneration.
3. We spend our earnings on products and behaviors that release more dopamine.

It is through the *meeting of needs* that the dopamine of point three above is released, generating within us the good feelings that are experienced when needs are met. Money is also transferrable, allowing the reward from one person's labor to move to another in exchange for the products of their skill, effort, and creativity. Step three could be as simple as purchasing a meal (especially a sweet one,) in which case the sensory

---

11. Sun, *The Fairness Instinct*.

response to the food in our mouth is rewarding, or it could be drinking a glass of wine. In that case there are two levels of reward: first from the flavor and other attributes of the wine and later the action of alcohol on receptors and ion channels in the brain. Alcohol has a complex profile with respect to neural activity, but one of its effects is to increase dopamine response by releasing endorphins.[12]

Alcohol as Proxy for Value

In writing this chapter I used phenomenology to explore the association between wine and money. Strange as it might sound, I found myself considering the monetary value of objects in terms of the reward I experience from drinking a couple of glasses of cheap wine. If the object had a value of ten dollars, I would ask myself "if I purchase this product, would the reward I receive from possessing, using or consuming it be equal to, less than or more than that received from a ten dollar bottle of wine"? This provided a benchmark I used to decide on the value of the commodity. The implication here is that when you pay for a bottle of wine you are paying for the dopamine-mediate reward derived from its consumption.

This principle carries over into the commodity market where certain neuro-modulating foodstuffs are used as benchmarks for value. These include Tea and Coffee (caffeine,) Cocoa (theophylline, theobromine, caffeine and phenethylamine,) and Sugar. Each of these commodities stimulate bio-chemical processes that contribute to dopamine release and an enhanced sense of well-being.[13] While the prices of commodities are anchored in demand (how desirable they are) they are also influenced by factors such as weather conditions, geopolitical events, and global economic trends, so the dopamine-value link is non-linear in practice. These products also possess other attributes such as flavor, texture, and smell that we find alluring.

In his book *Value(s) Building a Better World for All*[14], the former head of the Bank of Canada (and current Canadian Prime Minister,) Mark Carney, emphasizes the importance of humane values in a vibrant and sustainable economy. He introduces his argument using a wine-based parable shared by Pope Francis at a meeting of policymakers, businesspeople,

---

12. Zalewska-Kaszubska and Czarnecka, "Deficit in Beta-Endorphin."
13. Volkow et al., "Caffeine Increases;" Jenkins, "Why Your Brain."
14. Carney, *Value(s)*, 3.

academics, labor leaders, and charity workers at the Vatican a few summers ago. After joining the attendees for lunch, Francis opined:

> Our meal will be accompanied by wine. Now, wine is many things. It has a bouquet, color and richness of taste that all complement the food. It has alcohol that can enliven the mind. Wine enriches all our senses. At the end of our feast, we will have grappa [an Italian spirit]. Grappa is one thing: alcohol. Grappa is wine distilled.

He continued:

> Humanity is many things—passionate, curious, rational, altruistic, creative, self-interested. But the market is one thing: self-interest. The market is humanity distilled. Your job is to turn the grappa back into wine, to turn the market back into humanity. This isn't theology. This is reality. This is truth.

Here Pope Francis equates the mind-enlivening effects of pure alcohol with the self-interested quest for money embodied in an unregulated market. To Carney, the alcohol of the parable represents value. Its other attributes, the bouquet, color, and flavor that enhance the dining experience, stand for the values that temper an individual's natural avaricious tendencies.

In the same way that alcohol intoxication narrows one's focus to immediate desires and impulses, self-interest can lead individuals to prioritize their own needs and desires without considering broader perspectives. Alcohol addiction can also symbolize an obsessive focus on self-interest, where individuals become addicted to the pursuit of personal gain at the expense of other values. Self-interest is vital in driving the individual to meet their needs, but a balanced approach is required to avoid the negative consequences of overindulgence. These consequences are a cost to the one seeking benefit but might also negatively affect others.

The Dangers Inherent in Unbridled Self-Interest

An organization's unbridled pursuit of financial gain through the exercise of risky schemes can bring about its downfall. In jurisdictions where corporations have the legal status of persons, this financial collapse can be seen as a metaphorical (and fatal) drug overdose. The avaricious and ill-advised tactics exercised by management in these cases (a breakdown of pre-frontal executive function) result in the ruin of the corporation (the body.) A balanced free-market economy provides a framework in which individuals contribute the products and services that flow from their own expertise in exchange for those they lack the expertise or resources to deliver. In the absence of equity, however, this relationship can quickly become self-destructive or parasitic. This is the case where profit is sought without a consideration of risk or where personal gain for the producer is maximized at a cost to the health and wellbeing of the consumer.

Of note at this point is a distinction, made in Non-Violent Communication, between a need in the true sense and a *strategy* used to meet it. We *need* neither alcohol nor money, rather we use them to meet needs in the true sense or to generate positive feelings when needs remain unmet. It is in the attachment to these strategies that addiction manifests. Alcohol, through its action on the GABA receptor, brings peace of mind (a true need) and relaxes the body. Through the production of endorphins, it generates feelings of connection (another true need) with others that might not otherwise be accessible, but it also provides a direct way of escaping the pain of unmet needs by artificially boosting the activities of the nucleus accumbens and suppressing the amygdala.

Metaphors are powerful tools, but each one stands alone and only has meaning within its own limited framework. When we analyze any parable beyond its intended frame of reference it loses its meaning. The parable presented by Pope Francis made a distinction between wine and liquor to differentiate values from value, but a complementary parable presented in Habakkuk chapter 2 uses wine directly in place of alcohol and its effects, Habakkuk 2:5 (ESV):

> Moreover, wine is a traitor,
> an arrogant man who is never at rest.
> His greed is as wide as Sheol;
> like death he has never enough.
> He gathers for himself all nations
> and collects as his own all peoples.

While many of the English translations follow this format, some, like the New Living Translation, Berean Standard Bible, and Good News Translation, substitute "wealth" for the word wine. The alternate meaning derives from the Masoretic texts (the primary textual foundation for the Hebrew Scriptures in Judaism) and some Dead Sea Scrolls. On closer observation we see the maxims of deception, constant pursuit, greed, insatiability, and universality as equally applicable to both wine and riches. The neurobiological thread that unifies these word pictures is the limbic drive to obtain reward.

Money as a Source of Power

This maxim is also reflected in the pursuit of power. Money is power in that it provides the agency to meet our needs, or (more likely) to implement the strategies to which we are attached. The amount of power at play depends on the size of the arena in which we engage, but the principles are the same. In the United States, politics turns money into power and influence. Whether a candidate's financial backing is derived from grass roots supporters, political action committees, corporations or rich donors, it provides the agency to pitch their vision. The electorate votes for the vision that best meets their own needs. A candidate with integrity tempers their own self-interest in deference to their values and seeks to meet the needs of their constituents but, as with alcohol and other drugs, the disinhibitory effects of power promote sensation-seeking behaviors.

In his book *Disloyal* Michael Cohen describes the exhilarating high he and his colleagues experienced during the 2016 US presidential campaign. He writes:

> I was snorting from the same metaphorical mountain of cocaine, like Tony Montana in Scarface, stoned out of my mind on power.

In this surprisingly honest statement, Cohen establishes a phenomenological link between the felt-sense experience of cocaine intoxication and that elicited through his engagement in politics. The cocaine-induced high is the purest expression of dopaminergic reward known to humanity. Cohen's symbolic language emphasizes the link between the behavioral high and the drug-induced one.

Here are some other examples of how dopamine and money are linked indirectly:

> A popular night spot might provide rewards to the patron in the form of prestige, an attractive environment, great music and quality food and drink, but there is usually another aesthetic dimension. The management will likely hire attractive young women and men as servers and will enforce a dress code that enhances their natural

beauty. They know that people will pay more for that kind of company because of the reward experienced in beholding an attractive form and countenance. The prices on the menu will likely be adjusted upward as a result. Economists call this principle *marginal value*.

The high price of street drugs reflects their contribution towards dopamine activity. The incentive created by the anticipation of this reward is what gets the user hooked, which brings us to sex workers, colloquially known as *hookers*—another high-cost high dopamine "product."

The value of luxury cars is based on the prestige of ownership and the comfort and safety of the driving experience, all primary needs in Maslow's Hierarchy.

The enjoyment of music is also dopamine-mediated, so people are willing to pay hundreds of dollars to attend music festivals and concerts.

Inequality and Surplus Value

The principle of equity illustrated in the behavior of Capuchin Monkeys is a key factor in the history of class struggles. The indignation experienced by humans and other primates in the absence of equity is primarily a function of the amygdala. This small almond-shaped organelle is an important component of the limbic system and plays a central role in promoting the feelings and behaviors expressed when needs are *not* met. In this case equality and justice are key needs that factor into the equation. The amygdala is the seat of emotions like fear, anxiety, anger, and sadness.[15]

Capitalism has always promoted inequality, and those who receive less for their labors have always been indignant towards those who receive more, a point that caught the attention of a certain Karl Marx. In his seminal work *Das Kapital*, Marx expounds on his twenty-year analysis of capitalism and its limitations. The basis of his conflict with capitalism stems from the inequality embodied in *surplus value*.[16] I'm not a Marxist, but I do respect his analyses from a rational perspective.

---

15. Wikipedia Contributors, "Amygdala."
16. Marx, "Capital Vol. I - Chapter Twenty-Four."

While the goal of this book is, in part, to examine the role of motivation in addiction, motivation as expressed in the workplace is more complex. Every function of the CNS is a fractal. We can obtain enlightening insights by performing a surface reading of its activity, or we can delve deeper into its neurobiological and psychological depths to reveal more nuance. For example, in his magnum opus *The Wealth of Nations* originally published in 1776, Adam Smith[17] promoted organizational economic success as a function of efficiency, an approach that fails to embrace the complexities of human psychology. Marx, on the other hand, understood that focusing on efficiency alone would alienate the work force, bringing about a decrease in motivation and by extension productivity.[18] These ideas are explored in the research of Dan Ariely, Professor of psychology and behavioral economics at Duke University. Ariely theorizes that performance and remuneration are only indirectly linked. Looking beyond the role of money and self-interest as the sole motivators of productivity, he identifies a sense of organizational connectedness and the provision of meaningful work as more influential in long-term productivity gains.[19] In the context of the workplace, these principles expounded by Marx and Ariely parallel the *values* in Pope Francis's parable of the wine and Grappa.

Returning to our consideration of surplus value, here is an example of how it is generated. A worker is employed by a delicatessen to make sandwiches. They are paid $12 per hour and it takes two minutes to make a sandwich, therefore the worker receives forty cents for each sandwich produced. The raw materials cost $1.60, so the total cost of a sandwich to the owner of the delicatessen is two dollars. However, the sandwich sells for four dollars meaning that the owner makes two dollars of profit on each sandwich. Looking at this through the Marxist lens, the act of making the sandwich produces $2.40 of value. Forty cents goes to the worker and two dollars goes to the owner. The owner is therefore making five times as much from each sandwich as the worker does. As Marx saw it, the worker is actually only being paid for twenty four seconds of the 120 seconds it takes to make the sandwich, and for the remaining ninety six seconds they are working for free. These ninety six seconds of unpaid labor and two dollars of profit received by the owner illustrate the surplus value incorporated in each sandwich.

17. Smith, *An Enquiry*.
18. Marx and Engels, "Economic and Philosophic Manuscripts."
19. Ariely, *Payoff*, 54-5.

Undoubtedly Marx would have more to say about his theories were he alive today, but they do form a reasonably comprehensive description of how capitalism operates. Of note is the way he frames the discussion. His claim is that for each forty cents worth of work invested by the laborer, the owner is *stealing* two dollars. This language appears designed to tap into the indignation of the working class to bring about the revolution he saw as a pre-requisite for change.

Dopamine's Role in the Concept of Utility

Marx could not speak directly to the role of dopamine, but he did discuss the role of money in satisfying the *wants* of the individual. He frames the buyer's perception of value in terms of a commodity's usefulness or *utility*. Neo-classical economics, in contrast, reframe utility as the enjoyment or satisfaction the product brings the owner.[20] This recast concept of utility closely parallels the phenomenology of dopamine release when anticipating or enjoying the consumption or ownership of a product. The subjective nature of value is especially clear at auction where the amount of the winning bid (its perceived worth) depends on who is present and how much they *like* the object (*liking* being a function of the nucleus accumbens and its dopaminergic cycles.) At auction, the bidder predicts how much reward they will receive from owning the item on display and is willing to pay an amount of money equal to the predicted reward. In the research, this process is called *dopamine prediction* and it is something we all do in everyday commercial transactions. Sometimes the item purchased meets our expectations, sometimes it exceeds them and at other times it falls short. Misjudging the amount of reward resulting from a transaction is known as the *dopamine prediction error*. If the purchased item falls short of the prediction, we are disappointed, if it exceeds our prediction we are overjoyed. Recent research indicates that reward and the concept of utility are actually one and the same:

> Thus, dopamine responses signal formal economic utility as the best characterized measure of reward value; the dopamine reward prediction error response is in fact a utility prediction error response. . .this is the first utility signal ever observed in

---

20. CFI Team, "Neoclassical Economics."

the brain, and to the economist it identifies a physical implementation of the artificial construct of utility.[21]

The process by which I identified the link between reward and economic value (an exploration of internal felt-sense states and their rational interpretation) is a perfect example of how I have been using phenomenology throughout this book. First, I observe the phenomenon subjectively (upper left quadrant,) generate a hypothesis that explains it in objective terms and then analyze the existing research (upper right quadrant) for the physical implementation. There is no need for me to conduct my own experiments. For me this builds understanding of the objective backdrop to my own subjective experiences and perceptions. Objectifying these gives meaning to my struggles and helps me to know my enemy.

Dopamine Withdrawal

The amount of reward experienced in any situation tends to decline with repetition and familiarity. As familiarity increases novelty decreases and the individual will experience dopamine withdrawal[22] unless the level of stimulation required to maintain dopamine flow is maintained. (Dopamine Withdrawal is Charles Lyell's term and is not part of recognized medical terminology.) This prompts an unrestrained individual to crave ever-increasing levels of stimulation and repudiate anything that threatens their current life conditions. Consider this scenario for example—a couple who regularly visited a high-end restaurant took great pleasure in receiving a match book, custom printed with their names, each time they dined. This became their expectation at that particular establishment—they felt entitled to this gift on each visit. One evening they were told on arrival that their match book was not available that day. Their response was one of indignation! They had no such expectation at other restaurants, no matter how renowned, and experienced no displeasure when the gift was lacking at those locations. The indignation they expressed was triggered by the divergence of the situation from what they had anticipated, a result of dopamine reward prediction error coding.[23]

The incentive side of the limbic loop, driven as it is by the *anticipation* of reward, is particularly susceptible to disappointment. Our diner's

---

21. Schultz, "Dopamine Reward Prediction Error," para. 1, para. 17.
22. Lyell, "Dopamine Withdrawal."
23. Schultz, "Dopamine Reward Prediction Error."

elevated dopamine levels on arrival dropped precipitously when their gift failed to materialize; this upset the delicate equilibrium between the nucleus accumbens and amygdala. With the good feelings prompted by the expected reward deprived, the unpleasant feelings produced by the amygdala flooded in to fill the void. Expectation (reward prediction) is like a unidirectional tide—it flows ahead with ease but abhors anything that might cause an ebb. What our diners experienced is, in essence, the process of *acute* dopamine withdrawal. This takes effect whenever needs go unmet or when the strategies we have implemented to meet those needs are foiled. Chronic dopamine withdrawal comes into play when neurophysiological adaptations have taken place in response to long-term drug use, or when the self-structures associated with safety, esteem, and identity needs are poorly developed or damaged.

The unconscious mantra of the nucleus accumbens is "this feels good, and I want more." More spacious accommodations, a higher salary, a more prestigious mode of transportation, a Patek Philippe on the wrist perhaps. With each acquisition the owner becomes a bigger version of their former self. This is the angle expressed by Randall Rush, the Edmonton Alberta native who won $50 million on Lotto Max in 2014. In a National Post article, Joseph Brean[24] writes concerning Rush:

> He said [the win] taught him that quick money amplifies character, including a person's flaws and weaknesses. Money "makes you a bigger person of who you are," he said.

Could this be the principle to which Jesus refers in the *Eye of the Needle* metaphor? Is the amplification of character flaws and weaknesses (the bigger person) the confounding factor that prevents the rich from squeezing through the straight and narrow way? Matthew 19:23–25 (NIV):

> Then Jesus said to his disciples, "Truly I tell you, it is hard for someone who is rich to enter the kingdom of heaven. Again I tell you, it is easier for a camel to go through the eye of a needle than for someone who is rich to enter the kingdom of God."

Most of us in developed countries should take pause in absolving ourselves of guilt in this regard. While our income may be low in comparison with those we consider rich, the average resident of North America is extremely wealthy when compared with the denizens of most developing countries, upon whose resources we depend to maintain our elevated life

---

24. Brean, "The Lottery Curse."

conditions. While our society's reliance on money is unavoidable, we can evade many afflictions if we limit our appetites and satisfy ourselves with what is *enough*, 1 Timothy 6:10 (NIV):

> For the love of money is a root of all kinds of evil. Some people, eager for money, have wandered from the faith and pierced themselves with many griefs.

Unfortunately, for the addict there is no easy solution or quick fix and no escape from dopamine withdrawal. Recovery from all addictions involves a cost-benefit analysis (a consideration of value and values) and a lowering of expectations. At Celebrate Recovery I met a man who expressed gratitude that his drug of choice had been crack. This was due to the speed with which he hit rock-bottom in life, something that might have taken years with other drugs. When I asked him what he did to fill his inner void during abstinence, he described his previous evening's activities (a quiet night in) which included meditating and reading. These are not activities that elicit the kind of high delivered by a crack pipe, but they can be enough to foster feelings of contentment if the person we are assumes more natural dimensions. The bigger the person we are, to use the language of Randall Rush, the larger the void we can contain.

The key to my friend's continued abstinence was a change in thinking (a renewing of his mind Romans 12:2.) Whether we are attached to money or something else, we all have internal resources we can use to ensure our needs are met in healthy ways (an internal locus of control) rather than relying on external factors. All being said, we would do well to avoid making money, or drugs, the higher power that rules over our lives.

Thematic Takeaways

Feelings: We plan our activities based on our prediction of the reward (dopamine response) promised by those activities. Divergence from these projected outcomes manifests as the dopamine prediction error. If the reward is denied, or is less than predicted, acute dopamine withdrawal occurs. This is experienced in feelings of fear, anger/aggression, anxiety or sadness through the activation of the amygdala.

Parkinson's drugs enhance dopamine related function in the CNS. This effect increases the urgency involved in seeking the rewarding feelings that flow from everyday activities. It also amplifies those feelings when needs are met. This promotes behaviors that are out of character and potentially destructive.

Needs: Money is a storage mechanism for the fruits of our labor. We use money to obtain the products and services that meet our needs and in so doing experience the good feelings that flow from meeting those needs. These feelings are a response to the utility signal (a dopaminergic process) and are the subjective component of formal economic utility.

Neural Activities: Dopamine is highly active in the limbic reward pathway. It is also the primary messenger involved in movement, through its activity in the Substantia Nigra. The dopamine agonist drugs used to treat Parkinson's disease are designed to increase activity in the Substantia Nigra and promote movement, but they also activate the anticipation/reward machinery of the limbic loop. These drugs bind with dopamine receptors on the post-synaptic surface of the receiving neuron causing it to fire. This has a similar subjective effect to the administration of Ritalin and cocaine (dopamine reuptake inhibitors) and the amphetamines (dopamine inducers) but by a different mechanism (dopamine agonism.) Pramipexole and Ropinirole act directly on the receptors, filling the role normally played by dopamine, rather than utilizing the body's own dopamine supply, as is the case with the other two classes of drugs. The action of these drugs commonly promotes participation in risky financial expenditures and other behaviors driven by limbic activity.

Relationship: Money is not a living entity, but it is something to which we relate. The relationship we have with money manifests in the same dopaminergic cycles that bind us to those we count as friends and lovers. Our relationship with money can be as dysfunctional as a co-dependency and embody the same allure as a substance of abuse.

What you can do now: Start using compassionate curiosity and interoception to explore your relationship with money. Can you identify your own felt-sense experience of value? Does the allure of value threaten your values?

9

Gambling

"Gambling is not about money; it is about the thrill of the risk."
—Jeanette Winterson

Risk-taking behaviors are common to all, but the amount of risk tolerated or sought varies widely between individuals. Risk-taking is implicit in the Zuckerman Sensation Seeking scale and manifests most prominently in Thrill and Adventure Seeking (TAS) and Disinhibition (DIS).[1] Activities that entail a hefty dose of risk elicit an adrenaline hit. Some find this aversive while for others adrenaline promotes a pleasure response. Sensation-seeking is a complex trait as sensation seekers may be highly sensitive people, thick-skinned, extroverts, introverts or fall into many other categories.

Risk as a Distraction or Analgesic

Risk-taking behavior is used by some as a strategy to overcome the negative feelings that stem from trauma, abandonment, social anxiety, and self-doubt. As a six-year-old elementary school student I learned that breaking the rules provided an escape from the dysphoric sensations associated with confinement and the lack of autonomy imposed on me by the institution. During class I sometimes hid in a closet with a friend and played with the dress-up clothes or played practical jokes on other students. During recess you might have found me climbing trees, putting

---

1. Zuckerman et al., "Sensation Seeking Scale."

acorns in the exhaust pipes of teacher's cars or engaging in minor acts of vandalism, like peeling the outer bark off one of the school's cherry trees. In that case I was told by the principal that as I had removed the bark all the way around the trunk, the tree was going to die. I went back to visit that tree thirty years later and it was fine.

My most audacious act was to take a reel of fencing wire and insert the ends into the contacts of a 240-volt power outlet. My plan was to electrify the floor of the school so that it would be closed for the day, and we could all go home. Thankfully the contacts I chose were the negative and the earth. Following this incident the principal, always a lover of hyperbole, announced to the whole school "I've been shot down behind Japanese lines, I spent a year in a prisoner of war camp, but I've never been so afraid as when I pulled those wires out"! At the age of seven, the cumulative weight of my (minor) misdeeds led to a stint outside the principal's office during which I wrote "I must not talk in class" in an exercise book for three whole weeks! These behaviors were likely the result of undiagnosed ADD.

While these actions elicited a transgressive thrill and helped build a sense of camaraderie with my social group, the outcome was almost always a bad one. The alternative, however, was to suffer in an interminable state of powerlessness and stew in the hurts and doubts that sullied the visage of my unformed self. You can imagine how these episodes affected the conversations at parent-teacher meetings. On one occasion the School Principal told my mother "I've met boys like Stuart before, one of them came at me with a knife and I broke his arm! He'll go to jail for sure. Think of all the terrible things he's done at home." (Remember, I was six years old.) My mother racked her brain, but the worst thing she could remember was my picking the loose paint off the wall in the bathroom. While my behaviors at school went against social norms, they were never violent, and I have never spent a day in jail. They were, however, early tools I had developed to avoid the disquieting feelings that accompanied me into adulthood.

Risk is one factor involved in the thrill of *gambling*. Each time I chose to transgress the law of the classroom I weighed the value of the escape I sought against the punishment that might result. In each case this was a gamble.

The Maybe Principle

Returning to Sapolsky's Capuchin monkey experiment from the previous chapter, a change in the format of the test produced a surprising result. Using the new protocol, the frequency of food distribution when the lever was pressed ten times was reduced by 50 percent, in other words, on average, every second time the monkey performed the task they would receive nothing. This introduced the idea of "maybe" into the situation, and now when the light turned on the dopamine levels rose more than twice as high! The Las Vegas associations are clear, as the allure of gambling is driven by the anticipation of the win, but according to the research this anticipation would carry far less weight if a win was guaranteed.

With a little thought we can see why the *maybe principle* is advantageous for mammalian survival. Take fishing for example. One would think that sitting in a chair staring at a float on a lake would epitomize boredom, but fishing has the potential to become addictive when taken to excess.[2] The dopamine-mediated engagement comes from the anticipation of the catch. As most mammals are hunter-gatherers by nature, this anticipation makes the hunt for food rewarding in its own right. The ultimate reward is the flavor of the catch and a satiated appetite, but this doesn't consciously factor into the enjoyment of the hunt itself, as is evidenced by the fact that sports fishers usually return their catch. Esteem needs are also met by the exercise of skill, as can be seen in the proverbial statement "it was *this* big," usually accompanied by two hands placed unusually far apart. The maybe principle is also a big player in the hunt for a mate, a promotion or when engaging in a lawsuit. Being a winner is the goal, but there are often more losers than winners and losing does *not* feel good.

The Maybe Principle in Politics

In the previous chapter we considered the cocaine-like high experienced by members of a presidential campaign team in the run-up to the 2016 election. Undoubtedly risk played a role in this response as winning is far from guaranteed, but the principle also applies to the electorate who *bet* on a winner. This perspective is highlighted by the concept of Horse Race

---

2. Griffiths and Auer, "Becoming Hooked?"

Journalism[3], where political campaigns are presented using polling data and public perception rather than by considering candidate policy.

In a democracy, each voter is allocated a single bet (a vote.) While the influence of one vote is small, the process of selecting and wagering on a particular candidate is endlessly engaging for some. For others, politics is of no interest at all, and a continuum of interest lies between these two extremes. The relationship between a voter and the candidate is based on shared values and the voter's hope that the candidate will enact policies that support their ideals or living conditions. Personality politics also plays a role, where the voter admires aspects of the candidate, especially those which they themselves lack.

In media coverage of recent elections, pundits have repeatedly described voters as "smart betters" if they choose a particular candidate. One also described the leading candidate as "the good coke." This indicates the level of reward they elicit in segments of the electorate and links the felt-sense experience of political engagement with the activities of the nucleus accumbens. A candidate's job on the campaign trail is to incentivize as many as possible to cast a vote in their direction and this is achieved using the promise of reward. However, it is the maybe principle that can lift engagement to extreme levels.

The Psychology of Gambling in Food Marketing

While junk food is lacking in nutritional value, it is not normally considered harmful when consumed in moderation. The producers of such goods know they have at their disposal a potent tool for extracting moderate amounts of money from people's wallets. They draw heavily on research in neuroscience and psychology when formulating new products and their presentations. In chapter 2, I introduced the mechanism by which the umami flavor is generated using the food additives monosodium glutamate, disodium inosinate, and disodium guanylate, but there is another flavor enhancer that has similar dopamine-stimulating effects—capsaicin, the active component of chili pepper extract which binds with TRPV1 (the capsaicin receptor).[4]

FritoLay, a subsidiary of PepsiCo, manufactures a range of products that includes Doritos, a corn chip snack that comes in dozens of

3. Matthews et al., "The Mediated Horserace."
4. Marinelli et al., "Activation of TRPV1."

varieties. They understand the importance of novelty in engagement, so they continually formulate new varieties of the product. Their product locator[5] lists fifty-six Doritos products (at the time of writing) many of which are on limited release, ensuring an inventory of ideas that can be tapped into to revive flagging interest in the market if needed. On a recent trip to the supermarket, I found that most Doritos flavors contain capsaicin, these included:

Sweet Chili Heat

Spicy Nacho

Flamin' Hot Nacho

Jalapeno & Cheddar

Dinamita Chili-Lime Burn

The packaging for each is emblazoned with pictures of flaming corn chips, except Dinamita Chili-Lime Burn, where each tightly coiled chip resembles a stick of Dynamite with a lit fuse protruding from the top. In addition to capsaicin these products also contain the three flavor enhancers mentioned above. To me, FritoLay's most creative innovation is Roulette Doritos. This is a version of their popular Nacho Cheese product where approximately 17 percent of the chips in each pack (the actual percentage is a trade secret) are a special hot and spicy flavor. Despite my decades-long association with spicy foods of all kinds, I have to say these specially formulated chips are extremely hot!

Roulette Doritos use concepts and imagery from the popular betting game found in most casinos. When consuming them you never know whether the next chip will give you a capsaicin-derived dopamine kick (a win) or not. The goal is to activate the maybe principle in the consumer and boost their anticipatory and reward-related dopamine levels. Frito-Lay ships more product by leveraging the neurophysiology of addiction, they are after all the company that coined the tag line "I bet you can't eat just one." Doritos (when eaten in moderation) do not, however, promote harm to the self or others in the same way other substances or behaviors do. In fact, capsaicin has a list of recognized health benefits.[6] Gambling with money, on the other hand, can be extremely destructive.

---

5. FritoLay, "Product Locator."

6. Juber, "Health Benefits of Capsaicin."

Gambling With Money

The type of risk-taking discussed earlier is one thing, but when the risk involves money, the allure is greatly amplified. Gambling with money is one of those activities that has great appeal for some and is one of the few behavioral addictions codified in the American Psychological Association's Diagnostic and Statistical Manual of Mental Disorders (DSM.) During one study in which subjects were asked "on a scale from one to ten, with ten being the most pleasurable experience you've ever had, how would you rate gambling"? some participants responded with a "ten." How could this activity have such a profound impact on a person?

Unlike monkeys, humans have created a culture that includes many symbols, values and the ideas of past, present and future and use money as a representation of *future* reward. Money stands for possibility, ease and security. In addition to meeting our physical and psychological needs when redeemed, we also anticipate the future reward it bestowed as though it were happening in the present. The brain is ineffectual at distinguishing between an imagined activity and the experience itself. Anybody who has fantasized about being with their intimate partner knows this to be true. The memory of a kiss holds within it the same reward as the kiss itself, and the imagination can project into the future with similar results. In the same way the thrill of gambling is, consciously or unconsciously, a projection of future met needs.

Esteem needs are met for the gambler, in the present, by viewing themselves as a winner. In the future context, the packrat principle might drive an individual to hoard money. In this case the value of the money might never be redeemed, as merely owning the potential of reward is reward in itself. A cache of cash also meets the safety and security needs of the hoarder. This is perhaps the most insidious principle of wealth ownership, as the pooling of wealth for no purpose other than personal gratification (self-interest) makes it unavailable to meet even the *physical* needs of others. Hoarding can also be a symptom of another addiction concept—chasing the dragon. This describes the fruitless pursuit of the high experienced early in addiction, which due to a loss of novelty and neural adaptations has been placed permanently out of reach. This state is summed up in the words of Ecclesiastes 1:2 (ESV):

> Vanity of vanities, says the Preacher, vanity of vanities!
> All is vanity.

Gambling and Neurological Stimulation.

Dopamine inducers and reuptake inhibitors are stimulants. This is because they promote signaling in a way that accelerates neural activity, leading to their street name *Speed*. This prevents the brain from entering the sleep state and promotes insomnia. Gambling has the same effect for some, using the same mechanism. I once worked with a man who occasionally left the office on a Friday night and headed for the Casino where he would gamble non-stop until Sunday night. The dopamine hit he experienced from gambling nullified his body's natural need for sleep, allowing him to gamble for a full forty-eight hours. As he would win about 70 percent of the time at Poker, money was never an issue. He always made it into work on Monday, but I can't speak to the toll this took on him physically and mentally.

Gambling as a Life-Breaker

For those sensation seekers with high disinhibition scores on the Zuckerman scale, losing does not deter risky behavior. Consider this testimonial from the popular website and social media channel Humans of New York:

> I didn't even start gambling until I was forty three. I thought I was mature, but I was as vulnerable as a child. I started going to the casino once or twice on the weekends. It was a social thing. I'd just play cards with my friends. But I had good luck in the beginning. I started to win. And I started to love it. I couldn't wait for the weekend. Soon I started to go during the week. I'd work the early shift, and I'd have all afternoon to play. I abandoned all my responsibilities. Once you start playing, you forget that you're hungry, you forget that you're thirsty, you even forget that you have a family. I lost the grocery money, the rent money, everything. Winning felt great. And losing made me need to win. I'd make up excuses every time I came home empty handed. I'd say that I was mugged, or that my work hadn't paid us that week. Eventually I had to sell my car. I lost our house. I lost my wife. We'd been together for twenty years. I just took for granted that she'd always be around. The only thing that I didn't lose was my daughters. They sat me down one day, and said: 'Dad, either quit gambling, or quit this house.' And I never played again. (Medellín, Colombia)[7]

---

7. Humans of New York, "Medellín, Colombia."

This is not an anomaly. As I write, a story in the news tells of the six-year prison sentence imposed on Alanis Morissette's manager for embezzling seven million dollars in royalties from the sale of her music. He used it to maintain a lavish lifestyle and feed his gambling addiction. Another example, much closer to home for me, is that of a leader in my religious community who took tens of thousands of dollars from a charity account and spent it in the casino. He has since repaid the entire amount, but his actions caused me to question the mentoring role he played in my life. The Humans Of New York account exemplifies the triumph of values over value. Having suffered the loss of most things of value in his life, he was able to project into the future and feel the pain that would flow from the loss of his daughters. This is an example of the cost-benefit analysis that enables some addicts to escape their dependence.

The vices described here are pathological expressions of unmet need, and an attempt to meet those needs gone awry. Instead of evaluating individuals based on their actions, my objective is to delve into these inclinations within ourselves, without judgement. The process of recovery is a progressive releasing of obsessive attachments and a gradual acceptance of those things that have real value for us. To embrace these authentic values, we must turn our focus away from dopamine's pleasure response and incentivize ourselves towards the pursuit of meaning. This might seem daunting at first glance as it appears to return us to the painful state we sought to escape in the first place. If the only way to accomplish this was to grit our teeth against the unresolved pains of our past, there would be few success stories. What we need are means by which those hurts are permanently addressed. I will discuss a few of these in chapters 15 and 16.

Pleasure Is Meaningless.

As discussed in chapter 3, King Solomon used his immense wisdom to probe the relationship between dopamine and the material world, exploring the idea of pleasure using alcohol, women, work, and material goods. It is enlightening that he found these things empty (Ecclesiastes 2:1-11). After employing his massive material resources in the analysis of hedonism, his concluding words in Ecclesiastes 12:13-14 (NIV) are:

> Now all has been heard; here is the conclusion of the matter:
> Fear God and keep his commandments, for this is the duty of all

> mankind. For God will bring every deed into judgment, including every hidden thing, whether it is good or evil.

The implication here is that when all earthly rewards are exhausted, in that they can have no lasting value, there remains a reward with eternal implications for those qualified, based on the fruits of their *spiritual* labor (or the righteousness that is imputed to them through faith.) The writer to the Hebrews frames the incentive-reward cycle in eternal terms (Hebrews 11:6 NIV):

> And without faith it is impossible to please God, because anyone who comes to him must believe that he exists and that he *rewards* those who earnestly seek him (my italics.)

The hollowness of dopamine stimulation identified by Solomon is revealed in how it can be summoned out of nowhere, given the right medication. Parkinson's drugs, for example, don't just predispose a person towards gambling but also increase their spending on *stuff*. They enhance what the Bible refers to as *covetousness*—showing a strong desire towards material possessions—which is identified in Christian tradition as greed, one of the seven deadly sins.[8] Further investigation reveals that antidepressant medications can also trigger compulsive shopping behavior.[9] I have first-hand experience with this, having been prescribed Prozac for an extended period. My penchant was for shirts, jackets, shoes, and hats. Buying them and wearing them felt good, so I did it as frequently as I could afford. This is a state I remember clearly but having been free from the drug for a few years these things hold little sway over me.

Gambling on the Money Markets

In *Das Kapital*, Marx describes two separate cycles within capitalism[10]:

1. In the retail sector, where a commodity (usually labor) is exchanged for money, and money is then exchanged for other commodities (C—M—C.)

---

8. Britannica, "Seven Deadly Sins."
9. Sandberg, "Antidepressants."
10. Marx, "Capital Vol. I - Chapter Four."

2. The other is in the domain of Capital alone. Here, money itself becomes a commodity that is exchanged for more money (M—C—M+.)

The second cycle exists, predominantly, within the stock market and currency exchanges, where money is invested in the hope of further accumulation. The operative word here is *hope*, which takes us back to the idea of *maybe* revealed in Sapolsky's Capuchin Monkey trials. Hope is, in essence, the anticipation of reward, but not all rewards hoped for are realized. The creation of money from money on the stock market, and through arbitrage in the currency exchanges, bears all the hallmarks of gambling. The risk, game-like engagement and possibility of a big win are features of both, therefore those engaging in these activities must also be susceptible to addiction. Marx identified this addiction as the tendency to accumulate, accumulate without limit.[11]

Charles Lyell references the propensity of the capitalist towards addiction as one of the greatest threats to human health, and the sustainability of society:

> Power/money/esteem addicts are, by far, the most destructive of all addicts because they will do anything to maintain dopamine flow. They abhor truth and ruthlessly lie, cheat, steal, bribe, corrupt, demean, persecute, attack, destroy, and/or crush. To make matters worse, their addictions provide them with the resources that make it possible to ignore, obfuscate, or eliminate any and all threats to their dopamine flow. It doesn't help that insatiable dopamine cravings keep power/money/esteem addicts scrambling for the degrees, positions, and power that allow them to define legalities, moralities, and addictions. In a nutshell—Power/money/esteem addictions are the reason our species is flirting with self-annihilation.[12]

The Robin Hood Principle in Reverse

Class struggle is the clash between competing dopamine streams—those of the empowered and the powerless. The limbic systems of those in positions of power have been hardened over time to a point where humility and compassion become foreign concepts. I'm not claiming to be superior to these people; we are all human. It is just that I am not

---

11. Marx, "Capital Vol. I - Chapter Twenty-Four."
12. Lyell, "Dopamine Profile."

(partly by choice) in a position to be tempted. As Sir John Dalberg-Acton famously wrote "power tends to corrupt, and absolute power corrupts absolutely." This comment addresses the power of Pope and King, but in the same paragraph he mentions the Pharisees, confirming that the same principles pre-date capitalism. Jesus saw the tendency of money to turn the human mind away from compassion and morality. Among the rich of Jesus's time were the teachers of the law and the Pharisees, about whom he said in Luke 20:46-47 (NIV):

> Beware of the teachers of the law. They like to walk around in flowing robes and love to be greeted with respect in the marketplaces and have the most important seats in the synagogues and the places of honor at banquets. They *devour widows' houses* and for a show make lengthy prayers. These men will be punished most severely (my italics.)

The teachers of the law were primarily esteem addicts but also had a significant problem with greed. The devouring of widow's houses shows the tendency of those in power to pillage the poor to give to the rich (themselves.) Not only does money bestow power on its possessor, but power (in the absence of values) also generates wealth. It is the inverse of the Robin Hood principle. This would be less insidious were it not for the fact that some who lack power adore the powerful. I'll illustrate how this adoration meets safety needs using a couple of anecdotes from my youth.

I attended Sunday School for years with a wild young lady who liked to socialize with those involved in petty organized crime. One Sunday morning after the church service she recounted a ride she had enjoyed, in a high-end car, the previous night. She was captivated by her discovery of a couple long guns under the back seat and effused about how *safe* she felt in the company of their owners. In contrast, another female friend who stumbled upon a magazine ad depicting a shirtless hunk expressed her disdain for muscular men due to the peril their power portended. One was risk averse; the other was not. Risk is "a chance taken," it's a gamble. My Sunday school friend was risking arrest as an accessory of crime. She was also in danger that the power these men wielded could be turned against her, given the right circumstances.

What was the solution to self-interest that Jesus proposed? After a young man approached him and asked "Teacher, what good thing must I do to get eternal life?" Jesus replied: Matthew 19:18-21 (NIV):

> "You shall not murder, you shall not commit adultery, you shall not steal, you shall not give false testimony, honor your father and mother, and love your neighbor as yourself." "All these I have kept," the young man said. "What do I still lack?" Jesus answered, "If you want to be perfect, go, sell your possessions and give to the poor, and you will have treasure in heaven. Then come, follow me." When the young man heard this, he went away sad, because he had great wealth.

Jesus was asking him to abandon his reliance on money to meet his safety and social needs, instead surrendering his control to a higher power. Jesus's words are an arrow piercing the heart of the limbic loop. Such an attachment to money as a source of security, in preference to a reliance on God, is an internal rejection of God. However, money and security are inseparably linked, and we can't escape this fact. What we can avoid is relying solely on money for our support and protection. This young man approached Jesus, confident in his own standing, anticipating the reward of praise (and the esteem of those present) that was to be bestowed on him. Instead, he left in a state of dopamine withdrawal because not only had his immediate reward been denied, but his long-term reliance on money was identified as a confounding factor in his eternal plans. Jesus's commandment was radical, and I have yet to meet anyone who has honored it. It is, in my opinion, a hyperbolic device with a much more general application. He is saying—if you limit or reduce the volume of your assets (and the psychic space they inhabit in your thinking) you will fit through the narrow entrance to the kingdom more easily, a place where you will enjoy far greater riches.

James, the brother of Jesus makes clear why the reliance on money for security is an illusion. In James 4:13–16 (NIV) he writes:

> Now listen, you who say, "Today or tomorrow we will go to this or that city, spend a year there, carry on business and make money." Why, you do not even know what will happen tomorrow. What is your life? You are a mist that appears for a little while and then vanishes. Instead, you ought to say, "If it is the Lord's will, we will live and do this or that." As it is, you boast in your arrogant schemes. All such boasting is evil.

Evidently the need for security is fundamentally important to a functional relationship with God. When security is removed from the passengers of a plummeting plane there is little that money can do, but almost without exception every one of them will be praying to God. Israel's addiction to

Idolatry was an attempt to influence the uncontrollable and generate predictability. Sacrifices were offered to the gods of the land in times of famine and when marauding armies threatened the nation, or as an attempt to make peace with the locals. Idolatry was used as a sedative and Israel was also addicted to the stimulants of partying and temple prostitution that accompanied it. They could have turned to their own God, but he expected sobriety, and this was a commitment they couldn't master.

Commodity Fetishism

Marx discussed material goods using the notion of *fetishism*.[13] While the term is most used in our culture to describe some form of sexual proclivity, the word fetish also has a more ancient meaning: An inanimate object worshiped for its supposed magical powers or because it is considered to be inhabited by a spirit.

It is hard to imagine a more direct description of an idol, yet Marx used it to describe *commodity* fetishism. In its nineteenth century form it described the relationship between society as a whole and the commodities it produces. There is a sense in which these play the same role in meeting needs for security, identity, and esteem that the idols of Bible times did. This perspective results in commodities assuming a life of their own, in the social context, in which they have control over us rather than the inverse.

An Eternal Reward

The results of the primate experiments discussed by Robert Sapolsky prompted him to compare the differing roles of anticipation in humans and other primates. He concluded that our ability to project anticipation into the distant future was a uniquely human capacity. This manifests in the length of time a human is willing to endure between performing the work and receiving the reward. Many people go to school and work hard to qualify for admission to university. Once admitted they work to attain their degree as a pre-requisite to the job of their dreams. Once gainfully employed they exert themselves to obtain a house, family, holidays abroad, and many other benefits. This postponement of the recompense

---

13. Marx, "Capital Vol. I - Chapter One."

for hard work is known as delayed (or deferred) gratification.[14] The time lag between work and reparation is usually measured in years or decades, but Sapolsky stresses that humans are the only species capable of anticipating a reward that is realized after death!

The Matthew nineteen passage above admonishes the young man to relinquish his earthly riches and give them to the poor. In return he was promised "treasure in heaven." This did not mean, however, that when the man regained consciousness after death he would be presented with a cache of gold and jewels. This metaphor is expanded on in the parable of the Rich man and the barns Luke 12:16-21 (NIV):

> And he told them this parable: "The ground of a certain rich man yielded an abundant harvest. He thought to himself, 'What shall I do? I have no place to store my crops.' "Then he said, 'This is what I'll do. I will tear down my barns and build bigger ones, and there I will store my surplus grain. And I'll say to myself, "You have plenty of grain laid up for many years. Take life easy; eat, drink and be merry."' "But God said to him, 'You fool! This very night your life will be demanded from you. Then who will get what you have prepared for yourself?' "This is how it will be with whoever stores up things for themselves but is not rich toward God."

While material goods might empower us to take it easy, eat, drink, and be merry, those rewards are contingent on being alive. Earthly riches powerfully meet the need for security in this life, but only in this life—you can't take them beyond the grave. We are therefore presented with a choice: meet our security needs on a temporary basis by working for material resources or meet them eternally through our relationship with God.

While I have worked for companies that wrote bouncing cheques, the fact that I have generally been paid for the work I accomplished is an undeniable fact. I can therefore assume that the security provided by financial resources will naturally flow from the work I undertake. My anticipation of reward is based on my faith that my employer will make good, and this faith is based on experience. This is not so in my relationship with God. I personally know nobody who has received their eternal reward, except Jesus. While the temporal mechanics of post-life experience are hidden from us, what I do know is that nobody in my circle of acquaintances has ever returned from beyond the grave to attest to the greatness of heavenly riches. The belief in reward after death is therefore

---

14. Miller, "What Is Delayed Gratification."

left to faith in the words of God. Perhaps it is this *maybe* that keeps me engaged and committed to recovery.

Thematic Takeaways

Feelings: We seek relief from feelings like agitation, restlessness, uneasiness and boredom when they arise. If these feelings are a response to a situation over which we have control, we have the option of removing ourselves from that environment. If we lack the ability to do so, however, and those feelings are intense long-term responses to our situation, we might seek to boost our dopamine levels by engaging in risk-taking behaviors.

Needs: Risk-taking promises an escape from aversive feelings and a chance to meet needs of all kinds, but this promise is illusory. In the long term, physiological, safety, relational, and esteem needs are all threatened by our engaging in risky behaviors.

Ascribing to money and material goods a leading role in meeting our security needs is tantamount to idolatry. Our eternal needs are of greater importance than those expressed in the here and now, so living with eternal riches in mind brings with it the *possibility* of far greater gain.

Neural Activities: Primate research has shown a far greater dopamine response when anticipating a reward that is granted randomly compared with one that is guaranteed. This principle promotes risk-taking behavior in humans as the allure of the reward is amplified by the risk of losing it. This principle manifests in the pursuit of any perceived benefit be it food, a mate, a job or a political candidate.

Relationship: For the risk-taker (as illustrated in our Colombian gambling example,) the value (utility) of relationship can outweigh that of the addictive resource to which they are attached. This gives healthy relationships with friends, family, or our higher power, the potency required to detach the psyche from its fixations.

What you can do now: Are you a risk taker? Has your risk taking been costly in terms of health, financial wellbeing, or relationships? Consider what aversive feelings you are trying to escape by engaging in these behaviors. Our tendency is to deny the existence of these feelings, so don't expect to find them easily. These feelings arise when needs are unmet.

Use your intuition to find the level of Maslow's hierarchy at which your lack exists. Revisit the needs list to further clarify what you lack (baynvc.org/list-of-needs/) and see what this reveals about your life so far.

10

Technology

> "Taking a month to cleanse yourself from screens as much as you physically and possibly can, not shooting for perfection but just shooting for connection. You might learn something that you didn't realize about the people who are closest to you."
> —Shanan Winters

I AVOIDED TEAM SPORTS in high school, but the magical word "technology" filled me with glee. Everything from the school's obsolete mini-computer to the space shuttle provided me with a constant source of excitement. In 1973 my parents gifted me with one of the first electronic calculators on the market. I was enraptured by those green LED numbers and would repeatedly perform calculations for the thrill of knowing I had a thinking machine in my hands. Even the smell of it was hypnotic. I purchased my first liquid-crystal digital watch in 1975 with proceeds from my paper route. It was amazing! An electrical device that kept time with a quartz-crystal accuracy and communicated with me using seven cell LCD numbers. My number one thrill though was derived from the twice-yearly airshows at Biggin Hill airfield. The bigger, faster, and louder the plane the better, unless it was a hovering jump-jet, in which case the fact it *wasn't* moving was mind boggling.

Smart Phone Addiction

It sounds trivial, but one of the worst days of my eleven yearlong life was when the Sunday School Prize giving fell on the same day as the Biggin Hill Airshow and my parents gave me no choice but to attend. The church was close enough to the airfield that I could hear the planes but could not see them. There was some serious dopamine withdrawal involved in that experience. In today's technological environment, however, kids are vulnerable to that kind of misery daily. A psychiatric nurse once told me that most of the teenagers he sees arrive at the emergency department are threatening suicide because their parents have taken away their smart phone. This is an order of magnitude worse than my airshow experience because the social identity of these kids is defined through the device and the applications it provides. A smart phone is not just a pretty and powerful status symbol; it also allows them to meet their esteem and social connection needs in many ways.

While kids in my day might be content with collecting stamps, they now collect Snapchat Streaks (sequences of daily communications with their friends.) We hung out with our friends at the park, but this can now be accomplished remotely using Instagram Hangouts. While these activities might be dopamine inducing, the opportunities for dopamine withdrawal are also manifold. Kids experience the Fear Of Missing Out (FOMO) when they see posts and pictures from their friends at events they are unable to attend, or worse are not invited to. They can be bullied twenty-four hours a day. They may be in bed when an entire online friend-drama unfolds after eleven pm, causing them to lose hours of sleep. If parents take away their phone, they might take some other device (an old phone belonging to a parent, a tablet or a laptop computer.) If all the devices are locked up overnight in a safe, they might secretly borrow a phone from a friend. They might even commandeer a device left at their house without telling the owner. These are the behaviors of an addict.

Data plans can be turned off. WIFI network access can be put on a schedule. Apps like Netsanity can be installed on their phone so that their parents can see and control what the child is doing online, but kids are wily. They learn how to turn off the management software, or find locations nearby with free WIFI, or steal passwords for private devices. None of this makes the job of parenting particularly easy. It does, however, show how important relationship and connection are to the limbic system.

It's not that the problem is limited to kids, either. Some parents also spend hours a day on their devices—playing games, reading, and browsing social media. This in turn presents a poor example to the children and disrupts communication and connection within the family. Even before the smart phone came into vogue, the Blackberry Curve earned the nickname "Crackberry" because its use was (metaphorically) as addictive as crack cocaine. Such devotion to relationship proxies, however, does consume much of the time that would otherwise be spent on meaningful activities.

Software Designers as Drug Pushers

While the principle of technology-based reward might be viewed by the public as incidental to the design of mobile applications, that's only because the industry has been cagey about its strategies and goals. Make no mistake, the software industry is well-aware of the link between dopamine-mediated reward and application design; the motivation behind this strategy is money. This approach has given rise to a new job designation—the neuro-economist—and has spawned several companies and departmental specialties that seek to maximize the addictive nature of their products.

A key example of this is the Silicon Valley startup Dopamine Labs[1], the brainchild of neuro-economist Dalton Combs and neuro-scientist Ramsay Brown. A major feature of their approach to software design is maintaining the fast pace of rewarding actions. This utilizes a feature of the addiction process observed consistently in the abuse of chemical agents: the faster the delivery mechanism and the more frequent the administration, the more potent the dependency becomes. This has been shown repeatedly in comparing the relative effects of smoking, injecting, insufflating (snorting) and ingesting drugs. Over the years it has become obvious that technology addiction is a problem. Nowhere is this stated more clearly than in an interview with former Facebook executive Chamath Palihapitiya where he stated:

> "The short-term, dopamine-driven feedback loops that we have created are destroying how society works: no civil discourse, no cooperation, misinformation, mistrust. And it's not

---

1. Schieber, "Meet the Tech Company."

an American problem. This is not about Russian ads. This is a global problem."[2]

To give them credit, Combs and Brown took this to heart when designing the Mindfulness app "Space." Space introduces delays in the app browsing experience aimed at breaking the dependency on the fast-paced dopamine feedback process. When initially released, Apple saw the effect this app would have on their bottom line and refused to make it available on their app store. This decision was later reversed[3], presumably because while Apple's number one goal is maximizing profits for their shareholders, they do have a social conscience.

Mindfulness as a Defense Against Addiction

How does Space work? As reported in Psychology Today[4], mindfulness (the practice of conscious awareness of an individual's own inner thoughts and sensory states) activates the medial pre-frontal cortex. Through the process of neuroplasticity, mindfulness practice results in the growth of brain regions associated with self-awareness, empathy, compassion, and reason. However, as with a slowly growing plant, this change takes place gradually. While a few seconds of delay might help loosen the grip a rewarding experience has on the psyche, making long-lasting changes to the balance of power between the limbic and pre-frontal brains takes dedication, effort, and commitment. These are not attributes that humans have in large supply.

Addicted to Coding

While using smart phone applications might be addictive, creating them is no less so. I'm currently writing an app that uses the Global Positioning System to track the performance of students during commercial vehicle driving examinations. The dopaminergic effects of this coding task are palpable. I have been getting up an hour early just so that I can go to work and get the dopamine hit I receive while coding. I have not set an alarm in weeks because the stimulation the anticipation provides wakes me hours before I would normally get up. This combination of

2. Wong, "Former Facebook Executive," para. 2.
3. Baer, "Apple has Reportedly."
4. Rock, "The Neuroscience of Mindfulness."

dopamine and cortisol activity (both of which enervate the nucleus accumbens) acts like a stimulant drug.

The influence this principle has on the economy is worthy of note. In years past, companies paid thousands of dollars to purchase a single enterprise software license. Now a large part of our economy runs on open-source operating systems, databases, geospatial, and cloud systems that are available without cost. Where does the money that was previously invested in this software infrastructure go? These savings cannot but affect the economy. Also in question is how these products are provided without cost when teams of developers have spent thousands of hours creating them? In short, they are willing to donate their time and energy to these projects because they enjoy it. Their payment is no longer measured in financial terms but rather in the strength of the dopamine response (the value or utility signal) experienced during the team engagement and coding process. This field of study is now known as Open-Source Economics.[5]

Technological Transformation and Internet Porn

One of the most problematic uses of technology for dopamine modulation comes in the form of Internet Porn. This medium cranks up the hormonally enhanced adult limbic system and turbo-charges it using a litany of novelty-inducing experiences. In the early 2000's it was common knowledge that the online porn industry was big business. Figures available at the time showed that up to 65 percent of the cash flowing into the online payment processor PayPal came from subscriptions to porn and gambling sites. The policy of allowing these sites to use PayPal's services was curtailed in 2003[6] due to government restrictions under the Patriot Act.[7] Two individuals that made a killing on the sale of PayPal to eBay in 2002 for $1.5 Billion are Elon Musk and Peter Thiel. That same year Elon Musk founded his space exploration firm SpaceX. A year later he joined the electric car manufacturing business Tesla Motors. Assessing the amount of value generated for PayPal by the sale of internet porn is, admittedly, speculative, but based on their balance sheet in 2002 it must be a sizable portion. In making this claim I am not associating

---

5. Lagace, "The Simple Economics."
6. CNET, "PayPal to Levy."
7. Forbes, "The Wages of Sin."

either of these individuals with the porn industry, but the numbers are what they are, and the world operates the way it does because of our innate motivations and drives. From this we could conclude that much of the money required to establish some of today's most influential and transformative technologies was derived from the dopamine cravings of millions of libidinous web surfers.

A proportion of the funds acquired from the sale of PayPal, also went to Peter Thiel. In 2004 he established Clarium Capital, a global macro hedge fund. He is also the face behind Palantir Technologies, a "big data" analytics company. In October 2016 he donated one million dollars to the Super PAC Make America Number 1.[8] This organization supported the election campaign of then-presidential candidate Donald Trump. Some of this money also went to Cambridge Analytica, the analytics firm that spawned the Facebook privacy scandal in 2018[9] and helped Trump win the presidency. All in all, the effects of seed funding, whatever its source, can empower the privileged in their quest to mold the world into their own image.

Internet Porn and Neural Adaptation

But what about the consumers of internet porn? In what ways does it compromise their relational and spiritual functioning? Researcher Gary Wilson has dedicated years of his life to answering this question. In his TED talk at TEDx Glasgow,[10] Wilson outlines his research into the effects of repeated novelty-driven exposure to explicit videos and images. He claims that long-term exposure to the fast-paced discovery process afforded by hyperlinks and multiple browser tabs results in a form of physical dependence, and as with the repeated administration of chemical agents it causes physiological adaptations within the limbic system. This inhibits engagement in normal sexual behavior with intimate partners. Both the desire and the ability to perform are severely diminished.

Researching the causes underlying this effect in males proved problematic. In any scientific study a control group is required as a baseline for normal function, but in the context of online porn use it proved impossible to find a control group. Every potential volunteer was found to be a

---

8. Schwartz, "Billionaire Peter Thiel."
9. Confessore, "Cambridge Analytica."
10. Pangambam, "Gary Wilson."

user of Internet porn! Eventually a cohort of men who had quit using due to the negative consequences was located and the research proceeded. In most cases normal function returned after a period of six to eight weeks and those that abstained for an extended period became control group candidates. Wilson states, categorically, that the desensitization experienced by these men is due to the overexpression of the gene-transcription factor Delta-FOSB. It is true that, in animal experiments, Delta-FOSB has been found at elevated levels in the limbic systems of animals that exhibit behaviors such as compulsive running. Whether we can claim that it is behind the loss of interest experienced by men in these studies is questionable. What we do know is that our bodies have been designed with a brake that is applied when the reward circuit is overstimulated. This results from any repeated activity whether physical, intellectual, chemical or sexual. The goal of the nucleus accumbens is to incentivize us to achieve certain goals. If a goal is reached too frequently, adaptations occur in the dopamine signaling process that de-incentivize the individual towards that activity and return the system to homeostasis. The fact that these men recovered normal function after a period of weeks shows that the brake can be applied and later released.

What seems clear is that the problem manifests itself in the operation of dopamine receptors. These are G-protein coupled metabotropic receptors, meaning that while they bind with a transmitter molecule outside the cell membrane, they also work with a second messenger molecule inside the cell, usually cAMP or cGMP. These cyclic nucleotides effect several changes within the cell, from increasing the flow of electrical charge across the cell membrane by opening ion channels (a function normally observed in neurons,) to the activation of the protein kinase cascade (an enzymatic process that has many functions in different cell types.) The yet unfinished job of researchers is to untangle this web of interactions.

Is Addiction a Genetic Disease?

A different line of research has explored the link between addiction and dopamine receptor polymorphism. The term refers to the expression of two or more versions of the same receptor in different individuals. The dopamine D4 receptor, for example, has a long-allele version and a short-allele version which differ subtly in function. This concept may explain why two people react in different ways following the administration of the same chemical agent. Ultimately this line of research may identify a link between genetic differences and personality types, the hypothesis being that such traits as introversion, extroversion, thinking-focused, feeling-focused, sensing, judging, and perceiving are the result of receptor polymorphisms.

The Societal and Relational Impacts of Smart Technologies

As covered in chapter 1, a primary role of dopamine, in concert with oxytocin, phenethylamine, and the endorphins, is the establishment and maintenance of relationships. In a 2023 article published in Psychology today, authors David and Erin Walsh[11] discuss the role that smart phone use plays in the relationships of intimate partners. It concludes that the reduction in opportunities for connection between partners that results from excessive screen time affects their relationships negatively. A Politico article[12] makes the point that the frequency of sexual intimacy between long-term partners in North America is decreasing. It offers several suggestions as to why this might be the case, but a major factor is mobile device use. A Stylecaster article[13] entitled *How Your Phone is Ruining Your Sex Life* makes an interesting reference to phones in the context of relationship:

> The relationship we develop with our phones are so powerful, that it can actually take away from the intimacy you should have with your partner.

It is noteworthy that the rapport we have with these inanimate objects is described in the article as a relationship, but this is fitting as they stimulate the same neural circuits involved in relationship building. In this context, the brain distinguishes between competing stimuli and chooses the one that produces the stronger incentive. When we skip out on intimacy because we are glued to our device, the device is providing the stronger influence. The potential effect this might have on fertility levels seems obvious. The need for relationship can also be satisfied by activities, hobbies or interests. When sharing a beer with a friend at university who was in the homestretch of completing his PhD he raved about his field of study and how rewarding he found his work. He did, however,

11. Walsh and Walsh, "When Phones."
12. Wilcox and Sturgeon, "Too Much Netflix."
13. Stebner, "How Your Phone Is."

bemoan the fact that he had lost interest in his girlfriend. In the same way that limbic overstimulation through porn use can put the brakes on the incentive process, so can any sufficiently engaging influence.

A reformed porn addict made the point that, when engaging sexually with images of women, a subject is creating a fabricated relationship with that person. In the absence of the technology required to create images, either electronically or in print, the only way to achieve that experience would be through physical presence. Outside of non-consensual, power-over, and paid relationships this would indicate that the person had chosen to surrender to the viewer in the most intimate way. This unconscious process involves the reasoning that she chose you, you are worthy of her attention, she desires you. This engages us, not only sexually, but on the level of esteem and connection needs. Perhaps self-esteem and connection are the primary motivators behind many forms of relational dysfunction. We are not so much evil, manipulative and self-serving in our quest for stimulation as we are lonely, needy, and lacking in self-regard.

Technology in Bible Times

The link between technology and relationships has a long and storied history. In Bible times technological prowess was expressed through artisanship. The creation of art, tools, structures, ornaments and weapons of war. The ultimate execution of this was in the crafting of idols, concerning which Psalm 135:15–18 (NIV) states:

> The idols of the nations are silver and gold,
> made by human hands.
> They have mouths, but cannot speak,
> eyes, but cannot see.
> They have ears, but cannot hear,
> nor is there breath in their mouths.
> Those who make them will be like them,
> and so will all who trust in them.

Idols are distinct from modern technologies (apart from the role technology plays in commodity fetishism) in that they represent invisible spiritual beings that influence events beyond the realm of personal control, usually in a particular geographical area. Their main allure was in meeting the safety and security needs of an entire culture and its proselytes. Psalm 135 points out, in very clear language, that these entities do not exist. No matter how valuable the materials from which they were constructed, or the skill and artistry involved in their fabrication, they were powerless inanimate objects and any trust placed in them was misguided. They were, however, a major feature of every culture at the time. The God of Israel was different, as Psalm 42:2 (NIV) laments:

> My soul thirsts for God, for the *living* God. When can I go and meet with God? (my italics)

Any accord with an inanimate object is a pseudo-relationship. Seeking safety, security, and connection through such means diverts from the true source of those qualities, it also puts the brakes on our ability to experience intimacy with God. How many of us live in a "conscience-seared" state (1 Timothy 2:2) where intimacy is calloused over by our own neural adaptations to quick and dirty stimuli? God is unique in the Divine ability to provide minute-by-minute support, protection, and comfort, and (unlike the self-destructive quick-fixes we choose) God's influence transcends the grave. And yet idolatry was the primary stumbling block for Israel in their relationship with God. Idols were more than just objects, they were societal constructs with their own attributes, traditions, and rituals. What they lacked were those inconvenient values-based moral conventions decreed by the living God. Instead, their culture encouraged reveling, alcohol abuse and temple prostitution.

Addicted to the Bang

Another influential effect of dopamine can be found in America's obsession with guns. As a teenager I was a pretty good shot with a rifle and a shotgun, so I decided to invest in the only firearm I was able to acquire, a .22 caliber air rifle. It might not have had the same kick as a .22 rifle, but it was similarly engaging. I purchased it while working in the Forestry industry in Scotland, the same day I began a week of leave at home back in London. At the railway station in Dumfries, I was overcome by the craving to hold it. I unsheathed it from its case and admired the perfect sheen of its curvaceous exterior. It was, I thought, beautiful. I couldn't wait to test my skills shooting cans in our suburban back yard.

In an article coauthored by Dr. James Olds, Director of the Krasnow Institute for Advanced Study at George Mason University, entitled

Addicted to Bang: The Neuroscience of the Gun[14] he, and co-author Steven Kotler, explore the allure of guns in dopaminergic terms. Guns affect dopamine signaling in many ways. They are status symbols; they denote a sense of personal power, and they contribute to an (often false) sense of safety and security. The urge for safety, and conversely the fear of external threats, is largely governed by the amygdala (a major component of the limbic system.) The amygdala is known colloquially in some circles as *the Guard Dog*[15] and governs our tendencies toward fear, tribalism, aggression, and territorialism. The suitability of this epithet can be seen in the dog who protects their home and maintains allegiance to the family pack. When the dog detects an intruder on its territory it can't wait to sink its teeth into the invader. A confounding limitation, however, is the need for the teeth to be in physical proximity to the intruder for the bite to occur and there are many obstacles to achieving this. From the guard dog's perspective, possessing a tooth that could be thrown at the infiltrator at high speed would be empowering as it would circumvent the need for a chase and overcome the intervening distance. This is a basic function that the gun affords.

Olds and Kotler outline how these attributes contribute to America's gun problem and focus on the current fealty shown towards semi-automatic weapons. Their assertion is that it is the frequency with which these guns can be discharged that makes them especially alluring. This bolsters our previous discussion on the effects of rapid and repeated dopamine release in establishing dependence. It also explains the attachment shown, by many, towards bump-stocks (devices that use the recoil of the gun to operate the trigger multiple times in quick succession.) There is no practical application for these devices outside their involvement in mass shootings, and their ability to stimulate the dopaminergic thrill of the bang. The military has much more effective rapid-fire technologies.

The First-Person Shooter Video Game

For those for whom these opportunities are out of reach, there is always the thrill of the first-person shooter computer game. While the addictive quality of games extends well beyond the bounds of this genre, it provides a perfect example of the limbic activation involved in territory

---

14. Kotler, "Addicted to Bang."
15. BBB Australia, "Meet Your Guard Dog."

expanding activity. Another article in Psychology today[16] describes an experiment performed by Allan Reiss and his coworkers at Stanford University. The article details the results of brain scans performed on subjects engaged in a simple video game. As described in the article, the game operated as follows:

> The video game involved a screen with a vertical dividing line and leftward-moving balls on the right-hand side, which the player could click to remove. When a ball hit the divider, it caused the divider to move slightly leftward, reducing the player's "territory" on the left-hand side of the screen. Conversely, for each second that the area near the divider was kept clear of balls, it would move rightward, gaining territory for the player. The only instruction given was, "Click on as many balls as possible." All players soon deduced the point of the game and adopted a click strategy that increased territory.

The results showed the activation of key components of the forebrain's pleasure circuit, primarily the nucleus accumbens and the amygdala. While both male and female subjects exhibited these effects, they were significantly stronger in men. There are indications that the pleasure responses in men illustrate their predilection towards the establishment and maintenance of territory. Additional PET studies on game playing illustrate the specific involvement of dopamine signaling in the nucleus accumbens under similar circumstances.

Another study entitled *The Neural Basis of Video Gaming*[17] identifies a paradoxical effect in the brains of pathological gambling addicts. Their pleasure centers were activated when they *lost* money. Other studies show this to be a consistent feature in those who play computer games compulsively. If losing is rewarding it follows that a subject will chase both the winning and the losing scenarios resulting in extended game playing sessions. The principle is often used in advertising materials for computer games that draw in the potential buyer using terms like "this game will keep you up all night." While compulsive game-playing is good for the game manufacturer, it is also at the root of a disturbing trend towards men playing video games instead of getting jobs.[18] While the American Psychiatric Association were unwilling to include game play

---

16. Linden, "Video Games."
17. Kühn et al., "The Neural Basis."
18. Swanson, "Study Finds Young Men."

as a recognized addiction when compiling 2013's DSM V, it can only be a matter of time before such a classification is established.

Designer Drugs

The diversity encountered among colleagues in the workplace never ceases to amaze me. I once worked with a man who, in a previous life, was a dealer of Marijuana and Ecstasy (MDMA.) He had a friend who worked as a chemical engineer in a large urban center nearby during the day, and manufactured MDMA during his time off. My friend played the role of salesman for the product at raves in our local community. We got to discussing the subjective effects of MDMA consumption and the culture surrounding it. The rave and Electronic Dance Music (EDM) scene has been described as a rebirth of 1960's flower power; indeed, its official manifesto[19] describes it in terms of ecstasy (the emotion and the drug) and love. The sense of oneness and belonging experienced at these events is accompanied by respectful physical contact, usually in the form of hugs.

This brings us to a consideration of the number one technology addiction—dependence on the products of chemical engineering. This subject was the life-long pursuit of Alexander Shulgin (PhD Biochemistry, University of California at Berkley.) Together with his wife Ann, he wrote two books *Phenethylamines I have Known and Loved*[20] (PIHKAL) and *Tryptamines I have Known and Loved*[21] (TIHKAL). As already mentioned, phenethylamine is a primary agent in the phenomenology of love and is the root compound of a startling number of therapeutic agents and drugs of abuse. Any drug that is built on the underlying structure of phenethylamine is referred to as a substituted phenethylamine. Table 10.1 outlines some of the better-known compounds in this class and their common functions/uses. Of these, only DOM, one of the first *designer drugs,* was first synthesized by Shulgin:

---

19. Living Art, "Living Art Productions."
20. Shulgin and Shulgin, *Pihkal.*
21. Shulgin and Shulgin, *Tihkal.*

Name	Use/Abuse
Dopamine	Neurotransmitter/vasoconstrictor
Epinephrine (Adrenaline)	Flight-or-fight hormone and neurotransmitter
Norepinephrine (Noradrenaline)	Flight-or-fight hormone and neurotransmitter
Phenylephrine	Decongestant/Iris dilator
Salbutamol	Bronchodilator
Amphetamine	Appetite suppressor/drug of abuse/narcolepsy and ADD treatment
Methamphetamine	Broncho dilator/appetite suppressor/drug of abuse/narcolepsy and ADD treatment
Phentermine	Appetite suppressor
Methylphenidate (Ritalin)	Narcolepsy and ADD treatment
Ephedrine/Pseudoephedrine	Decongestants
Bupropion/Zyban/Wellbutrin	Anti-depressant/Smoking cessation aid
Fenfluramine	Appetite suppressor
MDMA (Ecstasy)	Drug of abuse
Mescaline	Drug of abuse
DOM	Drug of abuse

Table 10.1 Well known Phenethylamines

The medical community has learned some hard lessons from the use of these compounds. Aside from their role in addiction some are known to cause serious health complications. Likely the most well-known example is the combination of Fenfluramine and Phentermine[22] marketed as the appetite suppressor Fen Phen. In a small number of subjects, this combination therapy causes an increase in the thickness of the heart muscle and arteries in the lungs resulting in pulmonary hypertension, a progressive and potentially fatal condition. For this reason, Fen Phen was withdrawn from the market in 1997. Many of those who never developed the full-blown syndrome suffered from lingering heart valve problems for years afterwards.

22. Bang et al., "Pulmonary Hypertension."

Bearing in mind the long-term risks of using some of these compounds, it's amazing that Shulgin survived to the age of eighty nine! His modus operandi was that of personal bioassay (the assessment of drug effects by self-administration.) He tested the subjective effects of over 230 different compounds that he personally synthesized in the lab and ingested. On at least one occasion he synthesized a substituted phenethylamine, self-administered, noted the phenomenological effects and determined it could be a useful adjunct to the psychological counselling process. On the strength of this single trial he provided it to a psychologist friend of his for use on his patients! This can hardly be viewed as adhering to the drug-schedule laws of the US Food and Drug Administration.

In 1997, six years after publishing PHIKAL, Shulgin's penchant for biochemical exploration led him to publish his follow-up work TIHKAL. This chronicled his exploration of the Substituted Tryptamines. These are compounds that act on the serotonin (5HT) receptor family. Table 10.2 outlines some of the most common examples and their common functions:

Name	Use/Abuse
Tryptamine	Amino-acid
5-Hydroxy-tryptophan	Dietary supplement/serotonin precursor
Serotonin	Neurotransmitter/vasoconstrictor
Dimethyltryptamine (DMT)	Hallucinogen
Melatonin	Circadian rhythm modulator
Psilocybin (Magic Mushroom)	Hallucinogen
Sumatriptan	Migraine and cluster headache medication
Zolmitriptan	Migraine medication
Yohimbine	Hallucinogen
LSD	Hallucinogen

Table 10.2 Well known Tryptamines

The Use of MDMA in Couples Counselling

In the 1970s Shulgin introduced MDMA to psychologists as a psychopharmaceutical aid to couples counseling due to its promotion of feelings of intimacy and compassion. MDMA affects the function of at least three neurotransmitters: serotonin, dopamine, and norepinephrine. In a Huffington Post article[23] Carolyn Gregoire discusses the research of Matthew Kirkpatrick, a professor of preventative medicine at the University of Southern California and his contributors. Regarding the study Kirkpatrick states:

> MDMA slowed perception of angry expressions, increased psychophysiological responses to happy expressions, and increased positive word use and perceptions of partner empathy and regard in a social interaction.[24]

FMRI research has revealed that after the administration of MDMA, blood flow to the ventromedial frontal cortex (a brain region associated with the regulation and inhibition of emotional responses to social stimuli) is increased. It also results in decreased blood flow to the amygdala, which plays a significant role in instigating the emotional reactions associated with anger and violence. In considering the role of methamphetamine in promoting aggression we might assume it is due to increased activity in the amygdala; however, research indicates[25] that the amygdalas of meth users were just as well regulated as those of the control group. What was obvious, however, is that meth use results in a reduction of activity in the pre-frontal cortex. It appears that methamphetamine-mediated aggression is less to do with arousing the guard dog, and more to do with a lack of activity in the areas associated with emotional restraint.

How Does Amphetamine Use Promote Aggressive Behavior?

The article cited above, entitled *Don't blame the amygdala for Meth users' aggression* discusses the distinction between emotional insight and emotional regulation. Both principles are governed by components of the pre-frontal cortex but differ in their relationship with the amygdala. Lack of emotional insight flows from diminished emotional perception; lack

---

23. Gregoire, "MDMA Changes."
24. Wardle and de Wit, "MDMA Alters Emotional."
25. Arehart-Treichel, "Don't Blame."

of emotional regulation is the inability to control impulses of which the subject is already consciously aware. In both these cases the influence of the pre-frontal brain is reduced, giving the amygdala free reign over the subject's aggressive behavior. The acute and chronic effects of crystal meth use flow from a disruption of neural equilibrium rather than a bolstering of any component. The amygdala promotes aggression in methamphetamine users due to a reduction in the influence of its tamer.

Scripture as a Promoter of Pre-Frontal Cortex Function

In the context of religious observance there is an enormous contrast between the thought processes and behaviors promoted by meth use and those identified as desirable from a faith perspective in the Bible. In 1 Corinthians 13:4-7 (NIV) Paul describes many functions of the pre-frontal cortex:

> Love is patient, love is kind. It does not envy, it does not boast, it is not proud. It does not dishonor others, it is not self-seeking, it is not easily angered, it keeps no record of wrongs. Love does not delight in evil but rejoices with the truth. It always protects, always trusts, always hopes, always perseveres.

It is important to note that the Greek word for love used in this passage is *agape*, translated elsewhere as *charity*. This is not the kind of love we feel for our intimate partner (eros) whose expression is the responsibility of the nucleus accumbens. Agape is the kind of love that manifests as goodwill toward the other. 2 Timothy 3:1-5 (NIV) Paul goes on to identify the role of the limbic brain with amazing precision:

> But mark this: There will be terrible times in the last days. People will be lovers of themselves, lovers of money, boastful, proud, abusive, disobedient to their parents, ungrateful, unholy, without love, unforgiving, slanderous, without self-control, brutal, not lovers of the good, treacherous, rash, conceited, lovers of pleasure rather than lovers of God—having a form of godliness but denying its power. Have nothing to do with such people.

The law of the Hebrew Bible (the Christian Old Testament) was first formulated to address these attributes of self-interest and promote a healthy society. God knows how the brain works; God did, after all, design it. To illuminate the dangers presented by the primitive brain God provided the Decalogue, also known as the Ten Commandments.

The first five prime the reader against the primary limbic influences of self-aggrandizement and self-reliance. These are the underpinnings of the "higher power" approach to addiction recovery. I have worded these as simple imperatives as they were presented to Moses:

1. No other God beside me.
2. No carved images.
3. Honor my name.
4. Rest on the seventh day.
5. Honor your parents.

The remaining five deal with specific society-confounding expressions of limbic activity:

6. No murder.
7. No adultery.
8. No stealing.
9. No false witness.
10. No greed.

The remainder of the law elaborates and expands on these basic concepts. It describes the punitive sanctions to be applied for each offense, the ethics of business and commerce, a moral framework governing sexual relationships, and much more. The ultimate lesson of the law, however, is that the limbic functions created by God to perpetuate the species and protect the individual are impossible to control sustainably, hence our need for a redeemer.

Limbic function as a societal imperative

We could consider limbic function using the automobile industry as a limited analogy. Cars are generally overpowered and capable of exceeding the speed limit in most situations. It is rare for an automobile manufacturer to incorporate a governor into their product to limit its performance or top speed. This is because the driver needs speed and power to reach cruising velocity rapidly, move the vehicle out of harm's way when necessary and to overtake safely. Governing the pace of the vehicle is the responsibility of the driver's pre-frontal cortex. Likewise, humanity's drives can easily exceed the bounds of what is acceptable. If these drives were easily countered, humanity would likely descend into a morass of mediocrity, indolence and ultimate extinction. It is the invincibility of the limbic drive that propels humanity forward.

While the law was first presented as a solution to the problem of limbic supremacy and brought with it the hope of reconciliation with God, its goal was to teach this lesson: you cannot, by your own strength and works, overcome the enmity between God and humanity introduced in Eden. The pre-frontal brain cannot consistently overcome the influence of the limbic loop. As Paul writes in Galatians 3:23-24 (NASB):

> But before faith came, we were kept in custody under the law, being shut up to the faith which was later to be revealed. Therefore the Law has become our tutor to lead us to Christ, so that we may be justified by faith.

The law teaches us why Christ came: to achieve what for us is impossible by our own attempts to follow rule and precept—acceptance by our higher power, free from guilt and shame.

My local jurisdiction recently decriminalized small amounts of several illegal drugs, including meth, to remove the stigma associated with possession. The intention of this legislative change was to encourage users to reach out for help, free from the fear of legal repercussion. In a similar way, Christ de-stigmatizes us in our own eyes, giving us confidence that no matter how many times we relapse there is still hope and we are still worthy of love.

What seems paradoxical is that while the attributes of love (goodwill) described in 1 Corinthians 13 are rooted in pre-frontal function, its felt-sense characteristics (the incentive to seek it out and the experience of its expression) are a function of the limbic region. Being made as we are, we would not be motivated to pursue anything (whether destructive or of eternal value) without receiving a payoff. This is the thesis presented by John Piper in his book *Desiring God: Meditations of a Christian Hedonist*.[26] Without the expression of joy and satisfaction in this life and the hope of an eternal future (the payoff,) our worship would prove empty and hollow. God provides us with the felt-sense rewards of joy, satisfaction, and fulfillment in response to our faith in the Divine eternal purpose and our commitment to God's calling.

---

26. Piper, *Desiring God*.

The higher principles of goodwill, caring, and meaning itself would be inconsequential without the innate drive to pursue them. Without the burst of limbic energy in response to our acts of moral and religious observance, they would occur to us as meaningless. This aspect of the human mind embodied in the primitive brain forms the foundation on which all our higher moral functions are built. The moral dimension effectively subsumes the animal aspects of our constitution. Without the limbic foundation, all the attributes promoted by Christ in his ministry, and expressed in the pre-frontal brain, would collapse like a house built on shifting sands. These are ideas that we will discuss in more depth and from a different perspective in the next chapter.

Thematic Takeaways

Neural Activities: Mindfulness meditation increases activity in the pre-frontal cortex, promoting self-awareness, empathy, compassion, and reason. Strengthening the PFC increases executive function, a powerful tool for managing the addictive drives of the limbic system. Methamphetamine reduces PFC activity creating a vicious cycle where the addiction itself disempowers those parts of the brain required to abstain from using. Alexander Shulgin used the neuromodulator phenethylamine (PEA—the love drug) as the basis for synthesizing and exploring the subjective effects of dozens of chemicals. Many of these chemicals (especially the amphetamines) promote feelings of relational connectedness by promoting dopamine signaling.

Relationship: Smart phones are hubs for relationship management that provide ample opportunity for dopamine release and withdrawal. The user may also treat the phone as a proxy for social connection and in so doing develop an emotional attachment with the device that has many features in common with relationship. This attachment can resemble a low-level addiction. The neurological, neurochemical, and higher-level mental structures that mediate relationship are highly susceptible to being hijacked by substances, objects, images, and behaviors. These can trigger neurophysiological and psychological transformations that impede healthy relational behavior between partners. The title of Shulgin's book, "Phenethylamines I have known and loved," illustrates the use of relational language to describe the effects of inanimate substances, a principle that is equally applicable to other technologies.

What you can do now: Consider these two approaches to dopamine management: the dopamine menu and the dopamine fast. The dopamine menu is a list of healthy, intentional activities that provide dopamine in a regulated way. It replaces destructive habits (like doom scrolling) with constructive ones. The dopamine fast is a temporary abstinence from stimulating activities that "resets" your reward system. Although the Jewish practice of Sabbath keeping is primarily a religious observance, it also serves the role of dopamine fast by enforcing a break from stimulating activities. The goal of these fasts is to reduce overstimulation and restore sensitivity to natural rewards. The dopamine menu is a good match for those looking to substitute habits rather than eliminate them. The dopamine fast is a better fit for those wanting to regain control over compulsions. Take a few minutes to investigate these two practices to see if they are a fit for you.

11

The Devil You Don't Know

> "For I do not do the good I want to do, but the evil I do not want to do—this I keep doing. Now if I do what I do not want to do, it is no longer I who do it, but it is sin living in me that does it." —Romans 7:19–20 (NIV)

I PREFER TO VIEW sin as warning light rather than an indicator of personal failure and moral culpability. The behaviors identified by God as sin are here to warn us that something is out of balance and needs our attention. Abstaining from sins is not the goal. The goal is to fill the psychic void that pains us, with life-serving solutions. Sins are our teachers. Sins are normally poor strategies for meeting needs, but some may be predisposed to such behaviors due to underlying physical, emotional, or neurochemical dysfunction.

Brunner Syndrome

Brunner Syndrome provides us with an extreme example of the latter. The activities of our brains are driven by the release of neurotransmitters and neuromodulators that are deactivated when no longer needed. Those that belong to a class of chemicals known as *monoamines*, which include dopamine, serotonin, and noradrenaline, are broken down by the monoamine oxidase enzymes. These metabolize monoamines into their inactive, oxidized forms which terminates the signal initiated by their release. The MAO enzymes occur naturally within the CNS and maintain balance in the monoamine system. If they stopped doing their job, dopaminergic, serotonergic, and adrenergic function would continue unabated resulting in severe emotional and behavioral dysfunctions. This is the principle that underlies Brunner syndrome.

Patients with Brunner Syndrome experience dysfunctional MAO enzyme activity due to a mutation in the MAO-A gene which renders their enzymes inactive. The activity of the messaging molecules that would normally have been neutralized by these enzymes persists, leading to a state of hyper-arousal. An entire Dutch family, consisting of fourteen males, was diagnosed with this condition in 1993, the symptoms being aggressiveness, hyper-sexuality, arson and vandalism.[1] Since then, the same syndrome has been found in two other families. Were these individuals responsible for their destructive and anti-social behaviors? They could attempt to control them, but their chances of success would be severely limited by their condition. When we struggle with our own behavior, we should spare a thought for these families for whom the task of self-regulation is an order of magnitude more difficult than for the general population. Some theological *and* behavioral questions to consider regarding these families:

> Is the sacrifice of Christ powerful enough to save such a person?
>
> Does God consider our individual challenges when assessing our spiritual service?
>
> Are individuals with Brunner Syndrome responsible for their actions?

Another variant of the MAO-A gene that promotes much less extreme behavior than Brunner is present in 33 percent of people.[2] This gene variant is known as the MAOA-L (low activity variant) and produces less of the enzyme, leading to higher levels of monoamine neurotransmitters in the brain. This variant has been colloquially dubbed the "warrior gene" because some studies have linked it to increased impulsivity, aggression, and violent behavior, especially in males who are exposed to adverse environments like childhood abuse or neglect. This might go some way towards explaining aggression in the general population but runs into the classic nature/nurture argument that has to this point proven to be intractably

---

1. Prah et al., "Brunner Syndrome."
2. McDermott et al., "Monoamine Oxidase."

complex. One thing is clear, however—some people have a harder time self-regulating than others and through no fault of their own.

Monoamine Oxidase Inhibitors

The importance of having a well-balanced MAO system is illustrated by the action of a class of drugs called Monoamine-Oxidase Inhibitors (MAOIs) which raise the levels of neurotransmitters by preventing their oxidation by MAOs.[3] This has a similar effect to that of the Brunner related gene mutation. MAOIs are an effective treatment for those with low levels of monoamines, such as those suffering from depression[4] because they extend the duration of a signal initiated by a particular stimulus. However, they are contraindicated for those with normal monoamine levels due to the overstimulation that results from their administration in those subjects.

Their effects are also exacerbated by various foods and drugs. The list of compounds to avoid while taking MAOI drugs is long and intimidating.[5] Eating something as innocuous as Grapefruit while on these drugs can have serious, even fatal, side-effects. These include hypertensive (high blood pressure) and hypotensive (low blood pressure) crises. Food interactions are generally due to the presence of tyramine (an adrenaline, noradrenaline, and dopamine releasing agent.) Tyramine, like the amphetamines, is a substituted phenethylamine which modulates neurotransmitter release by binding with the Trace Amine Associated Receptor 1 (TAAR1.) Tyramine's effects are largely limited to the PNS, largely because it does not easily cross the BBB, so it has little effect on mood. The MAOI diet plan contains a list of thirteen different food groups and details how to minimize the negative effects resulting from their coadministration with MAOIs. This emphasizes how little control we have over our minds and bodies when our biochemistry is out of balance.

---

3. Cesura, "Monoamine Oxidase."
4. Mayo Clinic, "Monoamine Oxidase."
5. Hall-Flavin, "Avoid the Combination."

Making Sense of a Divine Paradox

One of the most challenging questions ever presented to me came from a young man I was mentoring through his preparation for church membership. The essence of his question was this: If God wants me to be a good person, and took the time to list dozens of things I need to refrain from to be regarded as good, why did he create me with all these desires, appetites and cravings to do exactly what he has outlawed?

God did not make the decrees of the law to stymie our enjoyment. They were put in place to address the negative actions of the unrestrained human brain, a brain which was lovingly instilled with drives that preserve and promote the survival of individuals, families and cultures. This is the dynamic I want to explore in more depth here.

The Chimp Paradox

In his 2011 book *The Chimp Paradox—The Mind Management Program to Help You Achieve Success, Confidence and Happiness*[6] Dr. Steven Peters describes the relationship between the angel on one shoulder and the devil on the other. The two are constantly at odds with one another. On a basic level we have two wills, and one is much stronger than the other, which helps to explain why relapse is an inescapable feature of addiction recovery.

Peters is a polymath, having studied mathematics, medicine, psychiatry, and performance sport. He served as a consultant Psychiatrist in Britain's National Health Service for twenty years. He also specializes in the fields of alcohol and addiction treatment. One of his greatest claims to fame is his work with cyclist (and now racing driver) Chris Hoy.[7] Hoy is an eleven-time world champion and six-time Olympic champion in cycling. With a total of seven Olympic medals, six gold and one silver, Hoy is the second most-decorated Olympic cyclist of all time. He credits his achievements in cycling to the system developed by Peters and described in The Chimp Paradox. The vision of the human struggle presented in The Chimp Paradox explains why, in the words of Romans 7, we humans do what we do not want to do.

The title of the book was chosen due to the strength differential between chimps and humans. While humans are predominantly

---

6. Peters, *The Chimp Paradox*.
7. Hoy, "Who Is Sir Chris."

land-dwellers, chimps spend their time in the trees, making their arms their primary means of locomotion. An initial study indicated that, in the task of pulling a weight horizontally across the ground, chimps could move five times as much weight as the human participants. More recent research[8] has shown that chimp muscles are only one and a half times stronger than those of humans, but the lesson remains—there are two principles at work within us, and one can easily overpower the other. The Chimp easily overpowers the human.

Peters' simplified model of cerebral activity divides the brain into three functional units:

1. The Human
2. The Computer
3. The Chimp

These describe three distinct attributes of the human brain, the prefrontal cortex, the memory, and the limbic system respectively. (It is useful to remember that the limbic system contains both the amygdala and the nucleus accumbens.) Chimps follow the law of the jungle, where survival is the primary goal. This manifests in fear/dominance and reward-seeking behaviors (as expressed by the amygdala and nucleus accumbens respectively.) Conversely, the human is expressed through the fundamentals of compassion, love, and reason.

Peters describes a comprehensive approach to managing the Chimp. Due to its superior strength, it cannot be overpowered, but it can be managed. We cannot deal with the more primitive aspects of our makeup using willpower alone. Peters frames this as follows:

> Having a Chimp is like owning a dog. You are not responsible for the nature of the dog, but you are responsible for managing it and keeping it well behaved.

He illustrates the relationship between the Chimp and the Human as a conversation between two brains:

> When you were in the womb, two different brains, the frontal (Human) and limbic (Chimp: an emotional machine,) developed independently and then introduced themselves to each other by forming connections. The problem is that they found they were not in agreement about most things. Either of these

---

8. Le Page, "Chimps Are Not."

two brains, or beings, could run your life for you, but they try to work together, and therein is the problem.

If the Human had full control, things would run smoothly in the moral sphere, but we would not accomplish much without being incentivized by the Chimp. The Chimp, on the other hand, is altogether too capable and operates without restraint. The Human needs a law as a guide on how to best manage the Chimp, but because the Chimp can easily overpower the Human this law is impossible to enforce fully.

This was not the perspective taken by the Jewish leaders at the time of Christ. They believed they were keeping the law, but Jesus revealed the folly of their assumption. Take this statement from Matthew 5:27–28 (NIV):

> You have heard that it was said, 'You shall not commit adultery.' But I tell you that anyone who looks at a woman lustfully has already committed adultery with her in his heart

The lesson for the legalist was that, despite the extreme constraints imposed by the letter of the law, the actual situation is far more challenging! He was telling them "You think you keep the law? You might be able to refrain from physical acts, but can you control your thoughts?" John makes a similar observation in 1 John 3:15 (NIV):

> Anyone who hates a brother or sister is a murderer, and you know that no murderer has eternal life residing in him.

Rather than establishing a stricter form of law in these statements, they emphasize the weakness of the law and its inability to save through perfect observance. These cases illustrate the futility involved in controlling the limbic mind, the seat of both lust and hate, but such is the fruitless task of the legalist. The law, through its ineffectiveness, nullified the concept of salvation by works. Christ, however, established faith as the real source of justification.

Reward Through Service

Despite these destructive attributes of the Chimp, it also holds within it the impetus to serve. In his book *Desiring God—Meditations of a Christian Hedonist*, John Piper reveals the inseparable link between desire and Godly service. In the same way that a healthy and sustainable society cannot exist when self-interest is the only foundation principle, our spiritual lives languish in the absence of self-interest. As the writer states in Hebrews 11:6 (NIV):

> And without faith it is impossible to please God, because anyone who comes to him must believe that he exists and that he rewards those who earnestly seek him.

We could define this reward as the endowment of eternal life on the believer in an afterlife, but Piper anchors it firmly in the present. To him, our lives are not built merely on the hope of a future eternity with God but also on the joy of worship and service right here and now. The same neural circuits that embody desire and reward of all kinds are vital to the connection we have with God and God's people in the present. The Psalmist frames it in the language of the appetites in Psalm 42:1-2 (NIV):

> As the deer pants for streams of water, so my soul pants for you, my God. My soul thirsts for God, for the living God.

Indulging this appetite is quite distinct from gratifying the senses through drunkenness and gluttony, or the administration of chemical surrogates for security and bliss. It is the gift of reward through giving rather than receiving, a reward Piper daringly describes as *hedonism*:

> I conclude from this meditation on the nature of worship that the revolt against hedonism has killed the spirit of worship in many churches and many hearts. The widespread notion that high moral acts must be free from self-interest is a great enemy of true worship. Worship is the highest moral act a human can perform, so the only basis and motivation for it that many people can conceive is the notion of morality as the disinterested performance of duty. But when worship is reduced to disinterested duty, it ceases to be worship. For worship is a feast.[9]

The chemical hedonist seeks to satiate their unmet needs through using, but as with the law, the ability of chemicals to provide long-term relief from the misery of a meaningless existence is limited. This is not so with the meaningful pursuit of a relationship with God. Piper continues:

> I do not come to the Bible with a hedonistic theory of moral justification. On the contrary, I find in the Bible a divine command to be a pleasure-seeker—that is, to forsake the two-bit, low-yield, short-term, never-satisfying, person-destroying, God-belittling pleasures of the world and to sell everything "with joy"

9. Piper, *Desiring God*, 97

(Matthew 13:44) in order to have the kingdom of heaven and thus "enter into the joy of your master" (Matthew 25:21, 23.) In short, I am a Christian Hedonist not for any philosophical or theoretical reason, but because God commands it (though He doesn't command that you use these labels!)

The Hedonism of Christian Love

We've discussed the similarity between the subjective effects of dopamine reinforcing drugs and that of loving and being loved. Regardless of the number of different Greek words for love we identify, each of which describes a different aspect of the emotion, I think we can all agree that love is at its core a rewarding experience. Jesus concisely pinpoints the value of love in the community of believers in John 13:34-35 (NIV):

> A new command I give you: Love one another. As I have loved you, so you must love one another. By this everyone will know that you are my disciples, if you love one another.

Can love truly exist between believers without being accompanied by a rewarding state, at least a sense of oneness? The verse seems to emphasize Johann Hari's claim that "the opposite of addiction is not sobriety, it's connection" and that to be an integral part of the body of Christ (1 Corinthians 12:27) is the pinnacle of connectedness. Jesus prayed to his father concerning his disciples in John 17:22 (NIV) using these words:

> I have given them the glory that you gave me, that they may be one as we are one.

If we, as Christ's followers, can be one with each other in the same way he is one with his Father, that is some intense connection! Lest we should question whether it is appropriate to receive a reward for loving our fellow believers, or whether this should be seen as a somber act of duty, I'll let John Piper do the explaining. In what is admittedly my use of a play on words, take note of his use of the scriptural word "gain" in this passage. In the context of loving one another it is appropriate that "gain" is the term used by researchers to describe an increase in dopamine signaling. Referencing 1 Corinthians 13, Piper writes:

> He [Paul] says, "If I give away all I have, and if I deliver up my body to be burned, but have not love, I gain nothing." If genuine love dare not set its sights on its own gain, isn't it strange

that Paul warns us that not having love will rob us of "gain"? But this is in fact what he says: "If you don't have real love, you won't have real gain." Someone, no doubt, will say that the gain is a sure result of genuine love, but if it is the motive of love, then love is not really love. In other words, it is good for God to reward acts of love, but it is not good for us to be drawn into love by the promise of reward. But if this is true, then why did Paul tell us in verse three that we would lose our reward if we were not really loving? If longing for the "gain" of loving ruins the moral value of love, it is very bad pedagogy to tell someone to be loving lest he lose his "gain." Giving Paul the benefit of the doubt, should we not rather say there is a kind of gain that is wrong to be motivated by (hence, "Love seeks not its own,") as well as a kind of gain that is right to be motivated by (hence, "If I do not have love, I gain nothing")?

The Chimp's Role in Addiction

The relationship between the Human and the Chimp is central to an understanding of addiction. Maintaining the conversation between the two is a vital task in countering the more reflexive influences of the latter. Left to its own devices the Chimp operates in the domain of reflex alone. The only way that balance can be infused into human behavior is through the involvement of the pre-frontal cortex, but as discussed by Gabor Maté in *In the Realm of Hungry Ghosts*, substance use degrades the connection between these two opposing parties.[10] The ability of the Human to manage the Chimp is diminished by this reduction in communication, giving the Chimp more of an influence on daily life and making it more difficult for the addict to refrain from using. This is the vicious cycle that makes recovery so challenging.

Maté believes this effect is the result of physiological damage. The drug of choice is seen as the cause of dysfunction in the lines of communication, either by damaging the pre-frontal region or the tissues that form the bridge between the pre-frontal cortex and the limbic system. I would add to this a second perspective—that the increased influence of the Chimp stems from a change in the balance of power.

I base this assertion on the following observation: the beneficial effects sought through substance use are experienced by the Chimp, and it is therefore within the Chimp that the most influential physiological

---

10. Maté, *In the Realm*, 168.

adaptations will take place (neurons that fire together wire together.) If the goal of using is to feel normal (or better than normal) by raising dopamine levels, then repeated use will raise the bar concerning the nature of normal and at the same time cause the limbic system to reassess how it responds to the new high. This might result in a reduction in the dopamine receptor count and/or adaptations in other structures that mediate limbic function. This has the effect of making the Chimp stronger or perhaps more desperate. It is now operating with a higher level of expectation (dopamine prediction) and at the same time its ability to meet its expectations have been diminished. The result is more anxiety, edginess, and craving when not using and less reward when using. The Chimp is now operating as a wild animal that has been backed into a corner and will likely attempt to satisfy its inflated cravings with little or no concern for those around it. This description of the Chimp, by Peters, is particularly germane to the Chimp in drug-withdrawal:

> As Chimps are constantly vigilant to danger, they tend to think catastrophically. They overreact to situations and fuel them with high and intense emotion. Whenever they perceive something is wrong, they have a tendency to start worrying about what might happen and then get things completely out of perspective. This frequently leads to terrible feelings of gloom and doom and stomach-churning moments[11]

As Victor Frankl notes in *Man's Search for Meaning*, with respect to the future, our brains are not limited to the anticipation of reward (nucleus accumbens); they also experience anticipatory anxiety or worry[12] (amygdala.)

Even a well-adjusted person is subject to another challenge from the Chimp—its speed of action.[13] Given a particular stimulus, the Chimp reacts four times faster than the Human, but thankfully the computer works twenty times faster than the Chimp and can assert itself before either the Chimp or the Human have a chance to respond. (These numbers are based on Functional Magnetic Resonance Imaging studies.) When challenged, both the Chimp and the Human look into the computer for information on which to determine their course of action. Initially they will likely be confronted by what Peters calls

---

11. Peters, *The Chimp Paradox*, 23.
12. Frankl, *Man's Search for Meaning*, 121–22.
13. Peters, *The Chimp Paradox*, 77–78.

Gremlins—negative elements of the overall belief system that have been constructed over a lifetime. Examples of Gremlins are "I can't handle life without using," "nobody likes me" or "the world is a dangerous place." These beliefs are embedded in the computer (memory) and can quickly influence the decision-making process of both Chimp and Human. The opposite of a Gremlin is an Autopilot. This is a helpful or constructive belief or behavior. Examples of autopilots include staying calm when things go wrong, focusing on solutions rather than problems and having a positive self-image. The primacy of the Chimp over the Human in functional terms informs the words of Paul that open this chapter, "it is no longer I who do it, but it is sin living in me that does it."

Over time we can evict our Gremlins and replace them with Autopilots. If our tendency is to panic when something goes awry but we have taken the time to develop a "don't panic" Autopilot, when the Chimp consults the Computer, it will see a big sign saying "no need to panic, you can handle this." A few milliseconds later the Human will see the same sign and indicate its agreement to the Chimp, at which point they can both embark on a more level-headed course of action. Autopilots constitute the self-structures that give strength and resilience to the psyche (spirit.) Building autopilots is the forte of the traditional worldview, a framework from which the higher-power approach to addiction recovery draws its power. No matter where your values fall on the worldview-zeitgeist continuum (more on that in chapters 16 and 17,) you'll find that putting aside judgment and embracing the traditional worldview as your foundation is an excellent first step.

Peters makes an interesting observation: "the Chimp isn't good or bad—it's a Chimp!" The limbic system isn't strictly bad. It is an emotional machine whose strategies and strengths are vital to our survival, both individually and communally. Any "bad" that comes from it is a result of the Chimp and Human not working together effectively. It is also clear that the Chimp embodies many positive aspects. Is anger always a bad thing (Jesus overturned the tables of the oppressive money changers)? The emotional aspects of love are rooted in the limbic loop (where the phenethylamines and oxytocin create attraction and commitment.) Is fear always a bad thing "Serve the LORD with fear and celebrate his rule with trembling" Psalm 2:11 (NIV)? Is joy bad "May the God of hope fill you with all joy and peace in believing" Romans 15:13 (ESV)? Claiming that the limbic system is bad is like saying that rewarding experiences and protective influences are always bad. The innocence of the Chimp is further evidenced by the fact that modern systems of justice do not consider animals to be capable of committing crimes. Criminal charges can only be applied to individuals who have the capacity for intent and moral responsibility, attributes that animals do not possess. As Peters emphasizes, the Chimp is like a dog in that it is without intent or morals. It does however have an owner, the Human, who bears responsibility for its behavior.

Although the Bible presents God as all-powerful, there are certain principles that God has either hard-wired into the cosmos by choice, or that God has abdicated power over. Take the parable of the wheat and tares for example. In this parable a landowner sows seed in his field, but an enemy comes at night and sows tares (weeds.) His servants ask "How then does it have tares?" to which the owner replies in Matthew 13:28-30 (NIV):

> 'An enemy has done this.' The servants said to him, 'Do you want us to go and gather them up?' 'No,' he answered, 'because while you are pulling the weeds, you may uproot the wheat with them. Let both grow together until the harvest.

Under the current dispensation, the good and necessary aspects of limbic function are so vital that the dangers of misuse are unavoidable. If you remove the "bad" from the Chimp it would cease to exist, along with its positive contributions. The Human creates balance by establishing an opposing tension with the Chimp based on values. As Peters puts it:

> The Human's agenda is to achieve self-fulfillment. This is usually about becoming the person you want to be and achieving the things you want to achieve. The Human will often search for the meaning of life. Many people might consider the Human as having the soul or spirit of the person.

To this I would add—the human has the conscience of the person.

The Chimp and Drug Withdrawal

Damaging the lines of communication between the Human and the Chimp is one of the negative consequences of drug use, but another becomes clear in the context of psychostimulant withdrawal. Dopamine withdrawal has both acute and chronic manifestations which result in a state of Emotional Lability.[14] This is a form of emotional dysregulation that bypasses both the prefrontal cortex and autopilots. In conditions like Bipolar Disorder, it shows itself in rapid swings between positive and negative emotional states, but during drug withdrawal the reactive states are uniformly negative. Lability has its root in the Latin word *labilis* meaning "to totter, sway, or move in an unstable manner." It is negatively expressed in uncontrollable bouts of sadness, irritability, and angry outbursts. Following a sudden break in the administration of dopamine-inducing drugs, neurochemical adaptations combine with the exhaustion of monoamine systems to create the purely negative emotions expressed by the amygdala. The feel-good phenomena promoted by the nucleus accumbens fail to materialize normally because it has adapted to an unnaturally high level of dopamine signaling and has consumed much of its neuro-chemical inventory. In addition to these withdrawal effects on the CNS side of the BBB, the adrenergic effects of psychostimulant drugs on many bodily systems, including mitochondrial energy production, lead to a state of profound bodily exhaustion which amplifies the sense of dysphoria.[15]

I experienced this phenomenological state following the discontinuation of both Ritalin and Dexedrine (dextro-amphetamine) administration as treatments for ADD. In this mode, the normal tensions of life can feel almost unbearable. The condition stems from a disturbance in the equilibrium between the nucleus accumbens and the amygdala, and a diminution in the influence of the PFC. This is a bird's eye view of the relationship between these organelles. In practice, the interactions between the PFC, nucleus accumbens and amygdala are so complex and enigmatic that we cannot claim to fully understand them.

In a healthy individual, the nucleus accumbens and amygdala work together to maintain a level mood. When both are equally active, a grounded state of being emerges. Higher activity in the nucleus accumbens may result in cravings and a state of anticipation or euphoria, but if amygdala activity exceeds that of the nucleus accumbens the individual experiences sadness, anxiety, rage or paranoia. If the two operate at a similar level (high or low) lability is suppressed. This makes alcohol an appropriate adjunct to states of both misery and celebration. The nucleus accumbens with its litany of "this feels good, and I want more" and the

14. Cuncic, "What Is Emotional Lability?"
15. Hartney, "The Comedown."

amygdala's "this feels terrible and I have to do something about it" are both answered by alcohol's effects. The former by alcohol's role as an endorphin inducer and the latter through GABA and Glutamate based suppression of the amygdala. In this context, however, we must remember the "angry drunk." A minority of individuals always buck the trend due to their genetic and epigenetic differences.

As the Chimp, with its superior strength and speed, subsumes both these organelles, psychostimulant withdrawal (high amygdala function and low nucleus accumbens function) introduces a confounding influence in the Human-Chimp relationship. In this case, the Human is trying to manage a Chimp in a highly agitated state, but without the contribution normally afforded by the computer—the custodian of the autopilots. (I speak to this through experience, without the support of formal research, this being another example where phenomenological awareness might inform the research community.) Here I am focusing on irritability.

At this point, words and actions become reflexive rather than measured. The behaviors and thoughts promoted by the amygdala take place with the knee-jerk certainty of a rubber hammer striking the patellar tendon. Awareness of these events emerges in retrospect, at which point feelings of regret arise, along with an opportunity to consider the phenomenological nature of flight in the absence of both pilot (PFC) and autopilot. This feature of withdrawal shows that relapse and withdrawal can be our teachers. Drug withdrawal can accelerate learning by increasing the frequency of the mistakes from which we learn, if we are open to it. Thankfully, as with opioids like Heroin, the initial withdrawal period of the amphetamines has a duration of between four and ten days, after which equilibrium is restored and access to the computer is re-established. Fortunately, amphetamine withdrawal does not carry with it the pain, nausea, and vomiting that accompany withdrawal from the opioids.

The periods between instances of relapse are our opportunities to invest in building new autopilots and strengthening those that exist. These form the self-structures and values the computer presents to the Chimp when it prescribes a harmful course of action. Complete abstinence from addictive behaviors can be achieved using a sufficiently comprehensive and robust set of autopilots. Once established, autopilots, with their superior speed of action when compared with the Chimp, can derail the Chimp's negative impacts before they enter conscious awareness.

Inducers Versus Reuptake Inhibitors

One thing that became clear from my breaks in drug treatment is that during Ritalin withdrawal partial autopilot function remains, but this is not the case with Dexedrine withdrawal. Why might this be so? Ritalin works primarily by blocking the dopamine active transporter in the pre-synaptic neuron, resulting in persistently high dopamine levels in the synaptic cleft. Any neural adaptation and functional depletion that takes place is consequently expressed in the down-stream neuron only. Dexedrine, on the other hand, binds with the Trace Amine Associated Receptor One (TAAR1,) an intra-cellular receptor found within the axon terminal of the pre-synaptic neuron. When bound, TAAR1 causes the efflux of spherical packages (vesicles) of dopamine through the axon terminal's membrane and into the synaptic cleft. At the same time, it reverses the flow direction of the dopamine active transporter, causing dopamine to pass through the DAT's central pore.[16] This adds to the dopamine released by the first process. The result is adaptation and functional depletion in both the up-stream and down-stream neurons. The experience of Amphetamine withdrawal, when compared with Ritalin withdrawal, may therefore be due to a more profound depletion of nucleus accumbens function, making the influence of the amygdala even more prominent. At least that's what it *feels* like. Add to that amphetamine's suppression of the Human (PFC) and you have a perfect storm, causing the psyche to totter, sway and move in an unstable manner.

Because the nucleus accumbens expresses the feelings experienced when needs are met, drug withdrawal, by reducing nucleus accumbens function, mimics a state of unmet need. This flips the coin on the situation where drugs induce an artificial state of satiation that fools the mind into believing that needs have been met. When you are in a state of drug withdrawal, your brain is unable to produce the feelings that flow from meeting needs in the normal way.

In addition to its expression during drug withdrawal, emotional lability is a function of reduced nucleus accumbens activity in daily life, showing itself in situations where needs go unmet. When we are hungry, for example, we might feel "hangry" (hungry and angry.) The following situations also promote a labile state and increase the activity of the Chimp:

---

16. Miller, "The Emerging Role."

Sleep deprivation

Thirst

Social Isolation (especially cabin fever)

Stress and overwhelm

Lack of physical activity

Unmet social or emotional needs

Pain or discomfort

We are all susceptible to lability from time to time. A Non-violent Communication practitioner shared such an instance with me. She had loaded the car with camping gear and kids and was heading out for a long weekend with her family, but something triggered her shortly after leaving, causing her to lose her temper. Her first reaction was to berate herself for allowing three years of NVC training to go down the drain in an instant, but she soon identified the cause—she had forgotten to eat breakfast. Her reaction was a hangry one. These outbursts are an unavoidable feature of the human condition and when they occur it is better to forgive and learn rather than judge ourselves or others.

Ease and Disease

Ease is a primary human need. It is the driving force behind the concept of convenience. When we accomplish tasks using labor-saving devices and strategies, we are freed from many of the unpleasant demands of life. When we are sick, however, we are in a state of dis-ease. Life is more difficult, and we are unable to engage in activities that meet needs under normal circumstances. Disease inhibits the nucleus accumbens allowing the amygdala to become ascendent. We describe the sick as *patients*, a term that emphasizes the influence of the Human over the Chimp. The initial attribute of love mentioned in 1 Corinthians 13 is patience, it is the first in the list of behaviors in that chapter that can be ascribed to the Pre-frontal Cortex. Being in a state of disease, especially when it involves pain or discomfort, begs for a higher level of patience, an especially important exhortation when a labile state is the default.

The Effect of Oxytocin on Amygdala Function

While the metaphorical Chimp is relatable due to our shared membership in the family *Hominidae*, the ultimate representative of law-of-the-jungle thinking is the king of the jungle, *Panthera Leo*. Rehabilitated lions are excellent research subjects in the study of oxytocin's behavioral effects due to the trauma they have experienced in captivity. These early life experiences predispose them to a chronic state of emotional lability. Jessica Burkhart, a Neurobehaviorist in the department of Ecology, Evolution, and Behavior at the University of Minnesota studies these animals in game reserves across South Africa.[17]

In a Zoom interview she described to me how she lures a lion to the enclosure fence using meat attached to a stick which she holds in place with her foot. When the lion bites into the meat it is temporarily unable to move away, allowing her to deliver a blast of oxytocin, intranasally, from a hand-held nebulizer. Two of her experiments illustrate the pro-social effects of oxytocin: a play object (pumpkin) trial and a roar playback trial. In the play object trial, the administration of oxytocin more than doubled the proximity tolerance of test subjects to other lions, when engaged in play, compared with the control group. The test subject allowed other lions to approach to a distance of three meters as opposed to seven meters for the control. In the roar trial, the occurrence of a roar response to pre-recorded lion vocalizations decreased from 50 percent of lions in the control group to 0 percent in the trial group.

She directly ascribes these behavioral differences to the positive effects of oxytocin on the nucleus accumbens and the associated inhibition of the amygdala this contributes. In addition to these acute effects, oxytocin appears to train the prefrontal cortex towards sustained pro-social behavior by developing long-term memories of the oxytocin felt-sense state. In short, in traumatized subjects, oxytocin reduces the occurrence of law-of-the-jungle responses to social stimuli. It follows that, through its action on the nucleus accumbens in humans, it would also reduce the drive to self-administer drugs of abuse to deal with trauma, abandonment, self-loathing and a host of other maladaptive cognitive behaviors. The research indicates that oxytocin administration does effectively reduce drug-seeking behavior in methamphetamine addicted rats.[18]

17. University of Minnesota, "Love on the."
18. Cox et al., "Oxytocin Acts."

Amygdala Suppression Through Social Connection

The amygdala can also be suppressed using exogenous drugs or by producing your own monoamines, including the endorphins and oxytocin. For a few years I manufactured these internally through my involvement in the Free Hugs movement.[19] This involved holding up a "Free Hugs" sign in a public space as an invitation for strangers to engage in a consensual embrace. Although I did this for the benefit of others, I also did it for myself. Through this activity I was managing one of my Gremlins—"you are not lovable." Each hug was a quick fix which validated my sense of worth, but with only fleeting effect. Over time I was able to transform this gremlin into the autopilot "You are as worthy of love and connection as others are." After forming this conviction, I no longer needed hugs to allay my anxiety.

From this I learned another valuable lesson—anxiety of this sort usually stems from our limiting beliefs (Gremlins.) When we use substances or external props to manage anxiety, we are treating the symptom and not the cause. Transforming the Gremlin into an Autopilot is the permanent fix. Instead of suppressing an agitated amygdala using stop-gap strategies, an effective autopilot prevents it from becoming distressed in the first place. For me, hugs are still a valuable contribution to others, but I no longer need them for self-validation. Seeking hugs for self-validation is a form of external locus of control—we are dependent on others to maintain our sense of wellbeing. The Autopilot provides emotional control with an internal locus.

In his book *In the Realm of Hungry Ghosts*[20] Gabor Maté recounts the words of a twenty-seven-year-old female heroin addict who described her first hit of heroin as a soft warm hug. Sarah McLachlan makes the connection between heroin and a loving embrace in the lyrics of her hit song *In the Arms of the Angel*. Here she describes the drug as an escape route from the pain of self-doubt (I am not enough) and the fear of endless exposure to a hard and dangerous world. But the song's title also frames the heroin high as a loving and protective embrace. Heroin fools the user's brain into believing that their safety and relationship needs have been met. Heroin and the endorphins produced by a warm embrace both bind with the mu opioid receptor with similar phenomenological effect, although, as with other exogenous drugs, the

---

19. Free Hugs Campaign, "Free Hugs Campaign."
20. Maté, *In The Realm*, 157.

effects of heroin are more pronounced than those of her endogenous cousins. The opioids soothe the fear and doubt expressed by the amygdala by increasing nucleus accumbens activity.

Oxytocin acts with similar effect but works directly on the nucleus accumbens. Besides hugging, another activity that raises oxytocin levels is eye gazing.[21] To engage in this activity, two people (whether an established pair or even complete strangers) face each other and maintain steady eye contact for a few minutes. This is the calming and connecting experience that spawned a global phenomenon—the Eye Gazing Movement.[22] Rounding out the list of amygdala-inhibiting social trends is the Cuddle Party Movement.[23] Cuddle parties are communication workshops that teach consent through physical contact. The first hour of each session establishes the ground rules that provide structure and security for the participants, allowing for two hours of peaceful connection through touch with whoever in the group consents. The workshop is overseen by a facilitator who ensures that the boundaries established by each participant are honored.

People are drawn to the Free Hugs, Eye Gazing and Cuddle Party movements by the feelings that result from their neuro-chemical influences. These are, in sum, the phenomenological expression of relationship. Having experienced a shot of intra-nasal oxytocin, on one occasion, I can speak to the similarity between the subjective effects of its exogenous and endogenous variants. The activities described above authentically meet relationship needs in ways that are only mimicked by drug use. But if these activities are used as quick fixes for our Gremlins, they will not lead to permanent healing. The popularity of these movements does, however, underline the important role that relationship plays in mental health.

Our Relationship with God

As the physicality of the external world is by nature an external factor and external influences are incapable of sustainably meeting needs due to the need for constant repetition, the ultimate relationship will have an internal locus, something that is fulfilled in a relationship with

---

21. Kerr et al., "Neurophysiology Of Human Touch."
22. Eckersley, "The Eyes Have It."
23. Darling, *Beyond Cuddle Party*.

Christ. While Christ's existence is external, our relationship with Christ is expressed internally and sustainably. As recorded in John chapter 17:22 (NIV), Jesus said:

> I have given them the glory that you gave me, that they may be one as we are one—I in them and you in me—so that they may be brought to complete unity. Then the world will know that you sent me and have loved them even as you have loved me.

And verse 26 (NIV):

> I have made you known to them, and will continue to make you known in order that the love you have for me may be in them and that I myself may be in them.

Setting aside the supernatural expression of the spirit of Christ in scripture, it is also materialized in our psyches. He embodies the values on which we build autopilots, giving us a robust defense against the Chimp. He knows our thoughts and actions, even those done in secret, making us accountable (accountability being a primary function of recovery groups.) We find union with Him through prayer, an *internal* conversation, giving us hope and allaying our fears. And it is through His love in us that the dysphoria of self-loathing is mollified and our lives find their ultimate meaning, revealing our worthiness as children of God. These are some ways in which the spirit of Christ manifests in us.

The emergence of this spirit does not come with the force of earthquake, wind or fire but as a still, small voice. Drugs work against this spirit, becoming a blight that rots its fruits of patience, kindness, gentleness, and self-control (Galatians 5:22-23.) But if we are open to Him, Christ soothes the feelings that flow from unmet need and becomes the rock on which we build our house. Not an empty house, or one filled with unclean spirits, but one filled with the spirit of Christ.

Thematic Takeaways

Feelings: The Christian hedonist appreciates the rewarding feelings that engaging in Christian service brings. These do not compare in intensity with the limbic phenomena produced by drugs but serve as a life-affirming alternative. After a period of adaptation, meaningful engagement in Christian service can permanently fill the empty house from which the demons of dependency have been expelled.

Needs: One of the most destructive gremlins is "nobody likes me," an indicator of poor self-esteem. This lack of esteem drives much of the dysphoria the addict attempts to escape by using. A healthy self-esteem autopilot enables the individual to meet their self-acceptance and self-connection needs and to matter to themselves (baynvc.org/list-of-needs/.) Without this autopilot the individual lacks faith in their own value and this prevents them from meeting their relational needs through others. The prime "other" in this context is their higher power; in the Christian context, the one who holds them in the highest esteem, no matter what.

Neural Activities: Our ability to moderate our behavior is severely compromised when our neurochemistry is out of balance. Brunner syndrome provides an extreme example of this, but other genetic or situational influences can also promote behavioral or mood disorders. Drug withdrawal is a prime example of this. The brain embodies a conversation between the prefrontal cortex and the limbic system. This conversation is enabled using structures in the intervening tissues, primarily the anterior cingulate cortex. Drug abuse can damage these lines of communication, making recovery more challenging. The limbic system also adapts to the unnaturally high dopamine levels promoted by drug use, making it more desperate and harder to control.

Understanding emotional equilibrium as a balance in activity between the nucleus accumbens and amygdala promotes self-forgiveness. This birds-eye view lets us depersonalize our reflexive behaviors and cravings and gives us an answer to the question "why." This question is a fundamental driver of human cognition, culture, and progress in the journey towards recovery.

Relationship: From the Christian perspective, the oneness believers have with each other, and with Christ, is the ultimate expression of relationship. The joy of relatedness is one of the greatest rewards we obtain from living in Christian community. When combined, the physical resources, recovery-support, boost in self-esteem, and security these human and divine relationships provide are what makes the higher-power approach to addiction recovery a successful one.

What you can do now: Our thoughts exert a powerful influence on which brain regions are active and determine which feelings we experience in the moment. If your thoughts activate your amygdala (perhaps by recalling traumatic memories) you will likely experience feelings of shame, anger, sadness or anxiety. Creating your own oxytocin can change the emotional landscape dramatically by inhibiting the amygdala. This might involve a hug from a safe person, spending some close time with a pet or even just thinking about them (a cognitive behavioral approach.) The CBT strategy is the most accessible approach because it has an internal locus of control, making you the boss of your feelings. By exercising your creativity, you will discover a collection of thoughts that bring about the desired transformation. The actual thoughts will be different for each person. The goal is to become aware of the moment the amygdala switches off and sense the positive feelings that emerge when it does. This is your proof that this strategy works and will identify which thoughts are effective healing elements for you.

12

Opioids and the Placebo Effect

> "Saint Augustine once said that God is always trying to give good things to us, but our hands are too full to receive them. If our hands are full, they are full of the things to which we are addicted. And not only our hands, but also our hearts, minds, and attention are clogged with addiction." —GERALD G. MAY

IN AN EARLIER CHAPTER we considered the vulnerability research of Brené Brown. For six years she wrestled with her belief in the dependability of analytical thought. Through this struggle she learned you cannot build a life on the foundation of objective rationality alone; the human experience is multifaceted and cannot be reduced to a dispassionate monologue or systems model. It took a year of counseling before she could accept the vulnerability of human existence, and the reality of the immaterial (that which cannot be measured.) She experienced the limitations of the lower right quadrant of the AQAL matrix and discovered the depth and nuance of the upper left. Being vulnerable we often need a means of support, but, as with Brown's experience, the support we choose may prove unreliable. In Isaiah 36:6 (NIV) the prophet warned of such a situation when Sennacherib king of Assyria laid siege to the fortified cities of Judah. During this quandary, he addressed these words to Hezekiah king of Judah:

> Look, I know you are depending on Egypt, that splintered reed of a staff, which pierces the hand of anyone who leans on it! Such is Pharaoh king of Egypt to all who depend on him.

Hezekiah was considering an alliance with Egypt that would save his people from the Assyrian armies, but Isaiah warns him to be wary. In situations like this it is tempting to choose the solution that provides the biggest bang for our buck in the short term, without considering the long-term consequences of our choice. Such is the predictable outcome of relying on addictions as a means of support. The outcome, as described by Isaiah in this passage, is further injury. Hezekiah heeded the advice of Isaiah and surrendered the situation to his higher power, the God of Israel, rather than exercising his autonomy and forging a political alliance with Egypt. In Isaiah chapter 37 this trust in the higher power pays off in spades and yet millions of people around the world turn to opium and her family for support during their challenging times.

The Cultural Roots of Opioid Use

Opioid use is deeply ingrained in our cultural consciousness. Historically they were a source of relief for the terminally ill who inhabited the opium dens of London and other cities in the late nineteenth century.[1] Opium was also the chain that bound China in servitude to the British in their role as the world's drug dealer. So important was the income from opium grown in Britain's Indian colonies to the national exchequer that two wars were instigated by Britain to protect the opium trade.[2] These wars eventually won them the territory of Hong Kong and exclusive access to many other Chinese ports. More recently opioids became the focus of addiction research, especially in the United States which with only 4.4 percent of the world's population consumes 80 percent of the worlds opioid supply.[3] This family of compounds includes such household names as Codeine, Morphine, Heroin, Fentanyl, Oxycontin (not to be confused with Oxytocin) and Vicodin, all of which are primarily used for pain management.

Opioids also play a vital role in the functioning of our minds and bodies. Like most drugs, the reason alkaloids extracted from the Opium Poppy, Papaver Somniferum, affect our state of being is that they mimic compounds already naturally present in our bodies and interact with a receptor network with profound reach. The name of the Opium Poppy

---

1. Gray, "American Opium Dens."
2. Hayes, "The Opium Wars."
3. Minnesota DOH, "Opioids Prescribing Practices."

species, Somniferum, is a Latin epithet meaning sleep-bringing, and has its pop cultural analog in the poppies of L. Frank Baum's book *The Wonderful Wizard of Oz*. These dazzling and fragrant blooms, when concentrated in sufficient number, induce a sleep in the unwary visitor from which they cannot awaken. Opioids extracted from the Opium Poppy are *exogenous* Morphines. Endogenous morphines (endorphins) are a family of neuropeptides (protein signaling molecules) that were identified in the 1970's by the American researcher Eric Simon.[4] He described an internal signaling network responsible for a plethora of objective and subjective effects. Gabor Maté describes the opioid signaling network as follows:

> Beyond their soothing properties, endorphins serve other functions essential to life. They're important regulators of the autonomic nervous system—the part that's not under our conscious control. They affect many organs in the body, from the brain and the heart to the intestines. They influence mood changes, physical activity and sleep and regulate blood pressure, heart rate, breathing, bowel movements, and body temperature. They even help to modulate our immune system. Endorphins are the chemical catalysts for our experience of key emotions that make human life, or any other mammalian life, possible. Most crucially, they enable the emotional bonding between mother and infant.[5]

He goes on to describe research conducted on mice whose opium receptors were deactivated by a gene targeting procedure. These mice were unable to attach emotionally to their mothers and lacked the ability to learn vital life skills because of the resulting relational deficits. In the previous chapter, we discussed the role of oxytocin in attachment and its role as a nucleus accumbens modulator. Considering the part oxytocin plays in the human response to hugs (earning it the labels love, cuddle and hug hormone) and its intimate relationship with dopamine, it makes sense that the endorphins would have a similar relationship with dopamine. Can we trace the similarities in the subjective experience of oxytocin and endorphins to a common root? Endorphins do indeed affect mood changes and produce a sensation of well-being by coopting the machinery of the limbic reward pathway. But unlike oxytocin, cocaine, and Ritalin they do this indirectly. Instead of attaching to receptors in the nucleus accumbens they bind with opioid receptors on the membranes

---

4. IES Brain Research "History."
5. Maté, *In the Realm*, 151.

of intermediary neurons, or inter-neurons and these activate the release of dopamine further down the signaling cascade. The opioids are another family of compounds whose rewarding effects vest in dopaminergic stimulation of the limbic system, underscoring the assertion that all addictions find their root in dopamine function.

The Phenomenology of Endorphin Binding

With practice and introspection, we can identify the specific reward experience (or felt sense) of an exogenous drug and begin to notice the same sensation in other aspects of our daily lives when that drug is absent from our system. These parallel experiences are brought about by the actions of the drug's endogenous counterpart and could be described as a "natural high." It might be in the afterglow of a blockbuster movie, reading a letter from a close family member, eating a delicious meal or when receiving a text message from our new flame. With an understanding of the full list of bodily effects elicited by an endogenous molecule it is possible to take a good guess at whether the reward you are feeling was triggered by adrenaline, serotonin, oxytocin or the endorphins. The feeling of well-being is your clue that the limbic system has been activated, and other bodily changes help refine your diagnosis. These non-drug experiences are not always subtle. Research has shown that when a teenage girl receives a text message from a boy with whom she is excitedly entangled, the amount of oxytocin disgorged is comparable with that released during orgasm! This helps explain why the combination of smart phones and social media is so addictive. It also illustrates the addictive quality of any non-drug experience or behavior that elicits a limbic response. An understanding of this principle is a necessary precursor to a twenty first century understanding of addiction.

For the remainder of this book the subject of self-esteem will be a frequent visitor. The amygdala mediated feelings stimulated by failure, self-loathing, and self-doubt are at the root of many addictions. The most profound example of this in my life highlights the role opioids play in relationship. I was raised by a loving family who did their best to give me comfort and security throughout my childhood. But, as with so many other children, the necessary corrections and reprimands my parents administered confused my inexperienced mind. How could my parents love me and be so mean? I went for the simplest answer to this question which was "I am unlovable," a conclusion I also drew from the behavior of my peers in High School. A few decades later, at a Sunday morning church service, the words of the pastor led me to the profound realization that I *was* loved by God. I visualized God's overarching presence and repeated the affirmation "I am lovable, and I am loved." As those words took form in my mind I was filled with a feeling of comfort and peace, but I also noticed something else. For the previous three months I had suffered from a persistent tickle in my throat and a dry cough, but as I held those words in my consciousness the tickle evaporated. After a minute or so it returned, so I repeated the affirmation again, and again the tickle dissipated. I soon I realized that the repeated affirmation had released neuropeptides that had suppressed the cough reflex and that this was likely endorphin related.

The tickle associated with a dry cough is processed in the brain in the same way as other sensory data creating a powerful urge to 'scratch' the itch. If you have ever been in a public auditorium when this reflex kicks in, you have likely experienced how impossible it is to control. Our cultural conditioning around remaining politely quiet and inconspicuous might motivate a speedy egress from the auditorium to avoid disturbing the proceedings. Activating the mu opioid receptor suppresses pain, but it also inactivates the cough reflex. This effect has been known for generations, hence the inclusion of codeine (a mu opioid receptor agonist) in medications for dry cough.

In the late nineteenth century, the German pharmaceutical company Bayer introduced Heroin to the market as a cough suppressant. A quick search online reveals dozens of ads, from the period, for "Beyer Heroin, the sedative for coughs." The most disturbing ads were those printed in Spanish which advocated Heroin as a cold remedy for babies.[6] It was freely available on pharmacy shelves until 1914, when it was designated as a prescription drug. Its use was banned by 1924 after the existence of widespread dependency had been confirmed.

Even users of codeine are at risk of becoming dependent, so cough-suppressing alternatives such as dextromethorphan[7] were developed that are less habit-forming. DM exerts its antitussive effect by binding with the Sigma-1 and Sigma-2 receptors which, due to this effect, were initially believed to be members of the opioid receptor family. This is no longer believed to be the case. DM is not exempt from abuse potential,

---

6. Edwards, "Yes, Bayer Promoted Heroin."
7. Erowid, "Erowid DXM."

but the risk is kept in check by the law of thirds. A third of people dislike its effects, a third are neutral, and a third enjoy its ability to improve mood and enhance the enjoyment of music, among other things. If you are a member of the last group, you might find it tempting. Attempts have been made to designate it as a prescription medication, but the overhead this would impose on the medical system would be significant. DM's indomitable market position is strengthened by the general belief that it is impossible to produce a cough suppressant with no abuse potential. In our city, there is an occasional spike in the number of young people admitted to the emergency department due to DM overdose, the usual symptom being respiratory depression. This has led some downtown pharmacies to remove DM based cough medicines from the shelves and keep them behind the counter.

The Role of Endorphins in Pain Reduction and Feelings of Love

So what of my "love" affirmation and its effects? I guessed that an endogenous compound had activated the mu opioid receptor system, and this was likely one or more of the endorphins. An online search revealed some papers confirming the antitussive action of the mu opioid agonist beta endorphin.[8] I raised this subject with my psychologist, and she agreed that the effect was likely due to the action of endorphins. I then asked, "if a pain killing drug can be released by repeating an affirmation of love, does that mean chronic pain can be treated using affirmations"? She nodded in agreement.

A search online for "Affirmations for chronic pain" reveals a long list of matches. The affirmations provided usually take forms like the following:

> I accept my life as it is
>
> I am stronger than my pain
>
> Life's challenges cannot defeat me
>
> My pain can never defeat me
>
> I have the power to heal
>
> My will to thrive is stronger than my illness

---

8. Kamei et al., "Antitussive Effect of β-Endorphin."

The Placebo Effect

My love affirmation, and these pain management strategies, all result in the conscious mobilization of the placebo effect. This involves the mind and body working together to bring about a significant change in bodily activities including immunological and autonomic function. In medical research this effect is due to the subject's belief that they are being treated with a pharmaceutical agent. Most medical research is conducted using double-blind placebo-controlled studies.[9] In these studies, the test subjects are divided into three groups. Group A receives the drug that is being assessed, group B receives a pharmacologically inert substance such as saline, sugar water or sugar pills and group C is a control group that receives no treatment at all. These studies are described as double-blind because neither the subjects, nor those who administer the treatment are aware of who is receiving the real drug and who is receiving the placebo. I'm not saying that the love of God, in my affirmation, is as fake as a sugar pill. What I am suggesting is that our structures of belief can stimulate real emotional and chemical changes in our brains and bodies. If we are given a pill that we believe will have a positive effect on our health, chances are it will. Furthermore, the effects of drugs are influenced, at least in part, by the placebo effect, in addition to their intended function. In most cases the subjects taking a placebo experience a significant improvement in their condition, but how does this work?

The emergency opioid overdose kits, available without prescriptions from Pharmacies in some countries, contain the mu opioid receptor *antagonist* Naloxone. This drug has a high affinity for the mu receptor and will inactivate receptors that are currently bound to Heroin, Fentanyl and other opioids. The effect of the drug causing the overdose is reversed and people that are experiencing respiratory failure will likely return to consciousness and lucidity within seconds or minutes. Fabrizio Benedetti of the University of Turin in Italy conducted an experiment on test subjects during which he inflicted pain on each of them daily throughout the study (don't try this at home!) He treated the pain with opioid pain killers, which were shown to be effective due to their action at the mu receptor. After a few days, he replaced the opioid injections with saline solution. Even though the liquid being administered was inert, the subjects still reported a significant reduction in pain following its administration. After a few days of treatment with the

---

9. David and Khandhar, "Double-Blind Study."

placebo, he added Naloxone to the saline solution.[10] Immediately the pain-relieving power of the saline was removed.

This indicates that the subjects were internally generating their own opioids through their belief that pain killers were being administered. The Naloxone in the saline blocked the action of the endorphins produced endogenously and a renewed perception of pain emerged. As I have shown, using anti-depressant, antihistamine, and cough suppressing drugs as examples, many drugs assert their action by binding with receptors at the cell membrane. These receptors are there by nature, so there must be an endogenous compound that is designed to bind at that site. If a placebo is administered, it is likely that the mere knowledge of the desired outcome spurs the brain and body to produce an endogenous ligand that mimics the perceived function of the drug.

It is regrettable that this effect has not been embraced by Western medicine as a starting point for the development of new therapies. This is likely because it is viewed by medical researchers as an impediment to progress rather than a tool to be exploited. In his book *The Faith of Biology and the Biology of Faith*, Robert Pollack describes the practice of removing the subjects with the strongest placebo response from double-blind studies, a process known as "placebo-washout."[11] This illustrates how threatening the placebo effect is to those who design drug trials, as it demands a much stronger response from the compound being assessed before it can be classified as an effective treatment. This in turn may classify a drug that cost millions to develop as ineffective.

Pollack also studies the role of the placebo effect in prayer. The belief that God has heard your requests and the requests of others concerning your health, is likely enough to trigger the activation of self-healing processes. Rather than seeing this as minimizing the role God plays in answering prayer, we could see it as a mechanism that Divinity instilled in each of us to promote healing. Pollack, through his foundation, The Center for the Study of Science and Religion (CSSR,) now known as the Research Cluster on Science and Subjectivity (RCSS,) hopes to initiate a series of meetings between those who believe in the existence of their immaterial soul and those for whom the existence of the soul is not so important as the reality of the placebo effect. Regardless of the perspective

---

10. Benedetti et al., "How Placebos Change."
11. Pollack, *The Faith of Biology*, 57.

taken, the placebo effect may prove to have been the most effective healthcare modality available in pre-industrialized societies.

Chronic Pain

Pain has many manifestations. Its purpose is to inform the subject of damage that has occurred, or is occurring, somewhere in the body. This is the experience of acute pain, but some individuals continue to experience pain at the injury site for years after healing has taken place. This is illustrated by the pain that occurs following a limb amputation, perhaps an arm, where pain might still be felt in the fingers. There is nothing there to hurt, in fact the neural structures that normally transmit pain from the affected area are completely missing. Other forms of chronic pain appear to have no cause at all. Western medicine is at a loss as to the treatment of these cases but does acknowledge the malady has a psychological component.

At the age of seventeen I attended the baptism of a woman in her mid-thirties who suffered from a chronic health condition that kept her housebound. Later, I had chats with her over tea about her faith, her love of wildlife TV shows and other subjects, including her pain. She had trouble walking and was often racked by pain as she lay in her bed, having to try and find a more comfortable position to lie in. Aside from the physical suffering she also had the added stress of uncertainty. The doctors could find no physical reason for her suffering. They did not deny she was in severe pain but saw her suffering as rooted in her psyche rather than the physical structures of her body. All they could do was prescribe Distalgesic, a combination of Tylenol/Paracetamol and the synthetic opioid Dextropropoxyphene, which did seem to ease her suffering. This pain-killing drug, which came to market in 1955, was banned in 2007 due to its perceived ineffectiveness, the number of overdose deaths attributable to its use and its negative effects on heart rhythm. Another drug, Oxycontin, experienced a similar fate in 2013 after it became the most abused drug in North America. People like my friend need better strategies for long-term pain management.

Opioids are effective at controlling pain in an acute setting, such as after surgery, but the risks involved in their use for chronic pain are vexing. Controlling chronic pain with affirmations (if that works for the individual) avoids these effects entirely. Like opioid drugs, affirmations can have an immediate effect on the perception of pain, albeit a more subtle one, but unlike the opioids their positive effects are magnified over time rather than being diminished.

Being Reconciled with Ourselves and God

The affirmations above deal directly with pain, but my affirmation "I am lovable, and I am loved" released endorphins due to their role in relational bonding. Prior to this, endorphin production was blocked by my attachment to perfectionism. A belief that perfect obedience is a necessary pre-requisite for a relationship with God renders attempts at building such a connection unsustainable. If you must perfectly follow God's laws to maintain a loving relationship, that relationship cannot stand. But Jesus nullified this pre-requisite by perpetually reconciling us with God through his sacrifice and execution at the hands of religious and state authorities, wherein he forgave his killers and was vindicated through the resurrection. His all-encompassing reconciliation was symbolized by the tearing of the veil between the holy place and the most holy place in the Jewish Temple at the time of his death. The barrier separating all humans from a direct relationship with God was removed. The atonement was complete, and an opportunity was provided for all to be at-one with God.

My own religious journey placed far too much emphasis on earning salvation through works and far too little on what Christ accomplished for humanity by giving his life. What I needed was to engage the relational machinery of my brain with God, and his son, rather than relying on external influences to control my emotional state. It was not until I began putting more emphasis on Christ as the doorway to a loving relationship with the Father that I was able to function spiritually again. This was the operation of the higher power on my amygdala and my opioid system. I had no worthiness of my own, but Christ gave it freely, enabling me to stand without shame in the presence of God. Only through that sense of worthiness was I healed. I am not saying it was a miracle, but it was the mind of God developing within me as he taught me through the failed solutions I had devised for myself and the suffering they brought.

A Challenging Commitment

It is not always easy to perceive the benefits of a relationship with God. Early in life I saw a commitment to God as toiling towards the unattainable while relinquishing all earthly reward in the process. Others spoke of their faith in God leading them to true happiness, while for me following Christ was a state of emotional subsistence, punctuated by using alcohol to dull the pain. The only strategies that made sense to me involved the consumption of external material and psychological inputs. The toiling towards such an unattainable goal leads to a meaningless existence.

This situation was exacerbated by my overly rational approach to the Bible. I was told that if you have doubts or questions about the Bible just read it and you will find your answers, but the more I read it the more questions I had! I tried to approach the Bible from a legal/scientific perspective using the principles of sound evidence-based reasoning, but this was a lower-right quadrant approach that ignored the subjective dimension. The solution for me was to stop seeing God as the punisher and to appreciate him as the loving father. This is not easy to do for those whose fathers were emotionally distant and the primary administrators of childhood discipline. Deciding to practice an active, daily communion with a God who is your biggest fan, rather than punisher, makes all the difference. It allows you to build a balanced picture of a divine reality in which endorphins, the building blocks of relationship, can flow freely. In a relationship where kindness and severity are factors (Romans 11:22) focusing wholly on severity is an impediment to authentic connection.

A Meaningful Life

A sense of meaning also flows from such a relationship with God. Victor Frankl dedicated his life to the study of meaning in mid-twentieth century Western culture. He described meaninglessness as a state of existential vacuum that was experienced by roughly 25 percent of his European students. In the United States, however, 60 percent of his students described their lives as meaningless.[12] He also referenced the research of Annemarie von Forstmeyer who noted that 90 percent of the alcoholics she studied suffered from abysmal feelings of meaninglessness.[13] He goes on to describe the research conducted by Stanley Krippner on drug addiction which revealed that 100 percent of the addicts he surveyed felt life was meaningless.[14]

---

12. Frankl, *Man's Search for Meaning*, 106.
13. Frankl, *The Unconscious God*, 97–100.
14. Frankl, *The Unheard Cry*, 26–28.

After decades of contemplating both the nature and expression of meaning in human lives, Frankl distilled it into this statement:

> [Meaning is] becoming aware of a possibility against the background of reality or, to express it in plain words, to become aware of what can be done about a given situation.[15]

These statements encapsulate what I would describe as hope, or the anticipation of reward—the primary function of the nucleus accumbens. The word *possibility* also echoes the *maybe principle* we discussed in the gambling context. In a more concrete statement, Frankl stated that the meaning of his life was to help others find meaning in their lives.[16] This has been and will remain a primary goal for me in my writing.

Frankl also proposed the concept of *paradoxical intention* which is built on two assumptions:

1. Fear brings about that which one is afraid of (e.g., the fear of not being able to fall asleep robs one of the ability to fall asleep.)
2. Hyper-intention makes impossible what one wishes (e.g. an artist or writer who obsesses over creating a masterpiece may find themselves unable to produce any work at all.)

In light of these two principles, I'll conclude this chapter with some questions for you to ponder. As a British ex-pat and Canadian citizen, it would be presumptuous of me to judge American culture as a whole, lacking as I am in the lived experience of being an American. For those of you who have the requisite qualifications to make such an assessment, however, I ask the following:

> Could a hyper-intention towards the *pursuit of happiness* actually make attaining happiness much more difficult?

> Do Americans tend to conflate happiness (serotonin) with pleasure (dopamine)?

> While America is a place where big concepts find their fulfillment, wealth inequality is higher than in most other countries. In Q4 of 2023 the richest 1 percent of US citizens possessed 16.6 percent of the wealth, while the poorest 50 percent had only 2.6 percent.[17]

---

15. Frankl, *Man's Search for Meaning*, 143–44.
16. Frankl, *Man's Search for Meaning*, 164.
17. Statista, "U.S. Wealth Distribution."

Does the expectation of living the American dream promote instances of dopamine prediction error coding (see chapter 8) in the bottom 50 percent?

Could this, in combination with the greed of big pharma, explain why the USA consumes 80 percent of the world's opioid supply?

Are opioid abusers using exogenous morphines (Oxycontin, Vicodin, Heroin) to replace the endorphins that are released naturally in those with meaningful lives, because their lives have no meaning?

Are the inevitable effects of using opioids to find happiness (anxiety, depression and mental health disorders) examples of paradoxical intention?

How might these principles apply to us all?

Thematic Takeaways

Feelings: The good feelings generated when needs are met are also accessible through affirmations. When we repeat an affirmation attesting to the fulfillment of a need and *believe* this need has been satisfied in our lives, our body and brain respond with positive feelings. This is the mechanism through which gratitude lifts our mood. It is a function of the placebo effect. Negative affirmations (negative self-talk, in this case a nocebo rather than a placebo) produce aversive states using the same mechanism even if the need in question has been met.

Needs: We all have connection needs, but we will not meet them if we see ourselves as unworthy of connection or if we judge those who are able to meet those needs as unwilling. These are both limiting beliefs rather than realities. Changing what we believe about connection, with God and other people, sets us free to pursue what had been denied us through our own doubt.

Neural Activities: Codeine and other opioids have been used for decades as cough suppressants. They achieve this by interrupting the cough response by binding with the mu opioid receptor. This reduction in sensory reactivity is also brought about by the release of endorphins, giving the observant human a way to detect endorphin activity and understand what is moving them in the present moment. A knowledge of the unique interoceptive profile of each neurotransmitter/hormone can reveal when they are active.

Relationship: The opioids, along with oxytocin, phenethylamine, dopamine, and serotonin are vital mediators in the formation and maintenance of relationship. The opioids are vital in creating the emotional bond between child and parent.

What you can do now: Search online for affirmations that address specific challenges you have in your recovery journey and explore their effectiveness for you. Bear in mind that the effect of these tools is generally cumulative, but for some people their utility declines with decreased novelty (the effect wears off with use.) Familiarize yourself with the interoceptive quality of these affirmations and explore ways of generating those same phenomena at will. You may find that with practice you can bypass the affirmation and directly generate your own endorphins when you need them.

13

Addicted to Love

> "I love you, but I want to love you enough that I never choose alcohol over you. Not even for a moment. I want to be someone you deserve." —Krista Ritchie

GOD IS THE ONLY entity that can share in our own subjective experiences. God knows what we think and feel, participating with us in the full gamut of conscious awareness. It is evident then that the kind of relationship we have with God is different from any other. It also differs due to God's intrinsic nature—immortal, invisible, omnipresent, and omnipotent—attributes that enable God to intervene in our times of need. A primary component of addiction is the need for security, and this is something Christ focuses on in Matthew 6:25–27 (NIV):

> Therefore I tell you, do not worry about your life, what you will eat or drink; or about your body, what you will wear. Is not life more than food, and the body more than clothes? Look at the birds of the air; they do not sow or reap or store away in barns, and yet your heavenly Father feeds them. Are you not much more valuable than they? Can any one of you by worrying add a single hour to your life?

Our relationship with God brings security, but without God-consciousness we are responsible for our own safety. In this we are limited due to the constraints imposed by our minds and mortal bodies. The comfort in Matthew 6 is that because we are valuable to God, and loved by him, he will provide for us in every way. Due to our being made in the image of God we experience love in the same way God does. In fact, the neurophysiological and neurochemical infrastructure that crackles with electricity within our skulls was designed by God to emulate him intellectually, spiritually, and emotionally. 2 Peter 1:4 (NIV) states that God:

> has given us his very great and precious promises, so that through them you may *participate in the divine nature*, having escaped the corruption in the world caused by evil desires.

When we turn to these "evil desires" or outside influences (drugs, elicit loves or money perhaps) for comfort and security, we are following Israel in their folly. The idols worshipped by the peoples of the Bible played a similar role to our objects of addiction. They also provided the perception of security and a sense of corporate belonging (lower-left quadrant.) It follows that the Old Testament history of Israel illustrates the same cycle of relapse and reform experienced by recovering addicts.

What we are addressing here is the proverbial God-shaped hole. The absence of God from our lives leaves a void. When we pour the objects of our addiction into this void it is filled temporarily, but the emotional/psychic rigidity they provide to our consciousness is short-lived. The process must be repeated *ad infinitum*, essentially a striving after wind.

The Supportive Properties of Air and Water

Let us examine this process using the ancient metaphors of air and water (according to Maslow, our two primary needs.) The inflatable Christmas snowmen seen dotted around town during the yuletide season provide a fitting illustration of the former. When driving home from work, which is after dark in these parts in December, they are upright and full of light. The impression they impart is one of energy, vigor, and jollity, but the faint hum of the pump maintaining their internal pressure alludes to the temporary nature of this state. On the way to work the next morning with the pump no longer running, they lie deflated on the lawn. Their apparent stability and sturdiness was an illusion created by the constant flow of air from the outside. They are much like the reveler who, the night before, was the life and soul of the party, but can barely get out of bed when morning comes. The issue with the snowman stems from his porous construction.

This is a frequent metaphor in contemporary English. When we are excited, we say, "I'm pumped"! and when we are low, we describe our state as "deflated." The deflated mood may be a result of too much alcohol the previous night, but it could also indicate a chronic emotional condition. Let us consider what contemporary culture has to say about the role drugs play in this inflation process.

The Role of Drug Use in Self-Inflation

In his song "Myself" the hip-hop artist Nav describes the addict's subjective experience of life. Copyright rules prevent me from including song lyrics in this book, but you can Google the song's title if you are interested in those. While the song illustrates some poor coping strategies it does reflect the experience of many and shows surprising perception and vulnerability on the part of the writer. It certainly was not written to glorify the abuse of over-the-counter drugs, but to depict a mind that exists in a persistent state of dysphoria from which escape is only possible by consuming externals. The chorus describes the protagonist's self-loathing and two strategies he uses to feel like himself again: consuming a "Four" and smoking Marijuana.

The active ingredient of Marijuana, THC, targets the canabinoid family of receptors and, as with the opioid receptors, these structures initiate a signaling cascade that terminates in the release of dopamine. A "Four" is an idiom for a mixed cocktail of promethazine, codeine and pop. Promethazine is a sedating antihistamine (histamine receptor antagonist) and codeine, as we have seen, is a mu opioid agonist. Both drugs reduce anxiety levels in a way that can transform a paranoid and aggressive person into a calm and gregarious one. Quite the off-label use for two medications designed to reduce mucus production and suppress the cough reflex.

The protagonist's goal is not to feel like a million bucks or feel like a god but to return to a balanced emotional state. He accomplishes this by suppressing amygdala function using exogenous substances. Once amygdala activity falls below that of the nucleus accumbens, the unpleasant feelings generated by the former are inhibited. We're not talking here about arrogance, but an inflation of the self to its natural size.

In his song "Righteous," another hip-hop artist, Juice Wrld, writes of the bottle of codeine cough syrup overflowing on the bedside table and the pills in his hand. His stated reason for abusing these medications is to fix the damage that has been inflicted on his psyche by past experience, damage that manifests in an anxiety of planetary dimensions. No doubt the song is autobiographical, as are the lyrics of another song, Legends, in which at the age of nineteen, he predicted his own death at twenty one, a prophecy fulfilled due to a fatal combination of codeine and oxycodone. Anxiety has many causes, but one of the most pernicious is the absence of a healthy self-concept and a lack of self-love.

The Dangers of Ego Over-Inflation

Although a healthy self-esteem is a pre-requisite to addiction recovery, over-inflating the ego can lead to other forms of addiction, giving rise to such conditions as anti-social personality disorder and malignant narcissism, both signs of an addiction to self-love. These maladaptive behaviors seek to heal the damages incurred early in life and escape the fear of these occurring again.[1] The narcissist accomplishes this by over-inflating their sense of status and value relative to their fellow man. This is accomplished by actively elevating self-perceptions and denigrating the standing of others, which provides a temporary (and destructive) escape from the fear and self-flagellation imposed by the amygdala. The "air" that inflates the narcissist comes in the form of external esteem (narcissistic supply,) something they actively seek from their associates. The apostle Paul uses the idea of ego inflation throughout his writings. Consider 1 Corinthians 13:4 (NASB):

> Love is patient, love is kind and is not jealous; love does not brag and is not arrogant.

The word arrogant is from the Greek word *Phuseoo*, a word derived from *Phusa*, meaning bellows. Although the Greeks lacked the technology to build inflatable snowmen, they did refine metals in a furnace using bellows to pump the air needed for a sufficiently hot burn. A smaller version of the same device was used to inflate the sheep's stomachs used in ancient bagpipes. The King James Version of the Bible translates the word "arrogant," in this verse, as "puffed up" to indicate the inflated image projected by those who are full of nothing but the "air" they derive from external sources.

God, in contrast, provides an inexhaustible source of emotional support. We are constantly in need of emotional refueling, especially

---

1. Ronningstam and Baskin-Sommers, "Fear and Decision-Making."

during hard times. 2 Corinthians 4:6–9 (KJV) describes this using the analogy of a clay vessel:

> For God, who commanded the light to shine out of darkness, hath shined in our hearts, to give the light of the knowledge of the glory of God in the face of Jesus Christ. But we have this treasure in earthen vessels, that the excellency of the power may be of God, and not of us. We are troubled on every side, yet not distressed; we are perplexed, but not in despair; Persecuted, but not forsaken; cast down, but not destroyed;

This passage emphasizes the ability of the knowledge of the glory of God to counter distress, perplexity, desperation, and hopelessness. Many writers have substituted the phrase "leaky vessels" (leaky containers) for "earthen vessels," illustrating our propensity to lose internal pressure over time. This holds true whether our contents are framed metaphorically as a liquid or a gas. Like leaky clay jugs our minds alone are unable to retain the metaphorical water (esteem, security, spiritual health, and encouragement) with which we fill them. Verses eight and nine of 2 Corinthians 4 emphasize the powerful benefits that come from the excellency of the power of God, something which offers the ultimate in security—respite from the fear of death. To be perplexed but not in despair is to have no way of escape, but to hold your own. The changes in emotional state brought on by promethazine and codeine are due to the suppression of the amygdala; God accomplishes the same result in us through faith. The relationship we have with God provides the structures that support us internally adding strength to our frame—Isaiah 58:11 (NIV):

> The LORD will guide you always; he will satisfy your needs in a sun-scorched land and will strengthen your frame. You will be like a well-watered garden, like a spring whose waters never fail.

This is the essence of the higher power—self-sufficiency through complete dependence. It is an interesting paradox, but one with a simple explanation. By depending on God, the one with whom we have a purely internal relationship, we are filled from within and become self-inflating snowmen and rigid plants rather than wilted ones. We are no longer dependent on externals, be they chemical, behavioral or those found in external reservoirs of esteem, but have an internal locus of control that flows from the deity. This renders our source of psychic nourishment inexhaustible. These benefits are initially possibilities rather than realities, but through a sincere desire for the things that are truly meaningful, they are attainable.

An Arid Environment

The Bible repeatedly uses the water metaphor which has special meaning for its immediate audience—the inhabitants of an arid land. Aside from air, our need for water is the most urgent of all needs and as the air filling an inflatable snowman provides internal rigidity, so water provides turgor to the leaves and stems of tender plants. There is no more dispiriting a sight to the eyes of the ardent gardener than a prized plant lying formless and wilted in the hot sun. As with earthen vessels, plants are porous containers of water. They have some control over water loss from the pores (stomata) in their leaves, but they still require regular liquid refreshment. If caught in time, a refreshing taste of water will return the plant to its former glory, but, in the natural world, the source of that water is invariably external. Isaiah 58:11 alludes to thirst to describe the feelings arising from unmet needs of all kinds.

God will not just satisfy our thirst but also our needs for love, companionship, and security. The verse presents images of a well-watered garden, one that is fruitful and flourishing. But notice how it concludes "*You* will be . . . like a spring whose waters never fail." God is drawing a contrast between the repeated administration of liquid sustenance and a constant flow welling up from within. He is contrasting an external locus of control with an internal one.

The Woman at the Well of Samaria

Christ used a similar metaphor with the woman at the well of Samaria from whom he had just requested a drink—John 4:9–18 (NIV):

> Therefore the Samaritan woman said to Him, "How is it that You, being a Jew, ask me for a drink since I am a Samaritan woman?" (For Jews have no dealings with Samaritans.) Jesus answered and said to her, "If you knew the gift of God, and who it is who says to you, 'Give Me a drink,' you would have asked Him, and He would have given you living water." She said to Him, "Sir, You have nothing to draw with and the well is deep; where then do You get that living water? You are not greater than our father Jacob, are You, who gave us the well, and drank

of it himself and his sons and his cattle?" Jesus answered and said to her, "Everyone who drinks of this water will thirst again; but whoever drinks of the water that I will give him shall never thirst; but the water that I will give him will become in him a well of water springing up to eternal life." The woman said to Him, "Sir, give me this water, so I will not be thirsty nor come all the way here to draw." He said to her, "Go, call your husband and come here." The woman answered and said, "I have no husband." Jesus said to her, "You have correctly said, 'I have no husband'; for you have had five husbands, and the one whom you now have is not your husband; this you have said truly.

Why was it that Jesus answered her request "Sir, give me this water, so I will not be thirsty nor come all the way here to draw" with a question "Go, call your husband and come here?" Could this be a reference to the addictive reliance the woman had on her relationships with men? (Regarding her marital status, we must consider that initiating a divorce was a lot more difficult for women than it was for men in New Testament times, so she was unlikely to have been the one who terminated her five marriages.) Aside from divorce, it is possible but unlikely that she had been widowed five times. Setting aside her marital history, it is clear she wanted to pursue a path of serial monogamy for the relational benefits it provided. Christ wanted her to see a parallel between her constant need for external reliance on water and her marital strategy. Both approaches involved the meeting of a need through the repetitive consumption of externals, but he offered her a better way.

His statement also showed that, while the search for love might be superficially alluring, its constant repetition is as monotonous and laborious as the daily drawing of water and its benefits equally fleeting. What Christ had to offer her was a spirit and system of belief able to meet her need for relationship, love, and meaning based on internal rather than external factors. He promised her a well of water within her springing up to eternal life, a meaningful and valuable existence that would add support and structure to her life when she wilted. The same need for repetitive action is also implicit in the Old Testament law through the cycle of sin, repentance, and sacrifice. But Christ's sacrifice removed the need for laborious repetition, being accomplished once and for all (Hebrews 10:1–4.)

Love and Addiction

In their seminal book Love and Addiction Stanton Peele and Archie Brodsky lay the foundation for a twenty first century understanding of addiction using a cognitive behavioral approach. The twentieth century medical community saw substances as the cause of addiction by direct action on the central nervous system (drug use predictably leads to physical dependency, but not necessarily addiction.) But in Peele and Brodsky's view, a view that is being progressively embraced, addiction is seen as a behavioral dysfunction that also operates in relationships. In the preface to the 2014 edition of their book they bemoaned the slow pace at which their ideas had been adopted by the Diagnostic and Statistical Manual of Mental Disorders. As we have already noted, the latest incarnation of the DSM (2013's DSM 5) only recognizes one behavioral addiction—gambling.

Their consideration of love may have application to the woman at the well of Samaria. Their elaborate and diligently researched thesis shows that addiction is, at its core, "any experience sufficiently safe, predictable and repetitive, to serve as a bulwark for a person's consciousness, allowing him an ever-present opportunity for escape and reassurance."[2] A bulwark is defined by Merriam Webster as "A strong support or protection"; in using this word they address the idea of a supporting structure for the consciousness. Through the release of endorphins, oxytocin, and phenethylamines, intimate relationships render the individual susceptible to addictive patterns of behavior. It is, of course, possible to engage in an intimate relationship without being addicted, but only if the relationship is based on a mature manifestation of love, which they state, "is possible only when we reach out to another person from our strengths rather than from our weaknesses."[3]

Alcohol and Opioids as Proxies for Love

The power of love to displace alcohol use was attested to by two of my acquaintances when recounting the early years of their marriages. They both lived dry lives for roughly seven years, but as endogenous endorphin production faded (a normal occurrence in relationships over time,) they found themselves returning to the bottle to reduce anxiety

---

2. Peele and Brodsky, *Love and Addiction*, 160.
3. Peele and Brodsky, *Love and Addiction*, 241.

and help them cope with stress. When alcohol replaces love, it is natural for the partner to see it as a rival, which in neurochemical terms it is. It is understandable, then, that unregulated alcohol use in the home, especially on the part of one partner, becomes a cause of conflict. Both love and alcohol, through the production of endorphins, inhibit the amygdala, allow the body to relax, and the mind to cease its frenetic activities. The apostle John attests to this in 1 John 4:18 (ESV) "perfect love casts out fear." These words, when applied to our relationship with our creator, are of special relevance.

The writers of the TV show "House" showed awareness of this principle when producing season six episode twenty-two of the series. Doctor House, a recovering Vicodin (hydrocodone) addict, had just lost a patient with whom he had a meaningful connection. On the same day, his long-time love interest, Doctor Cuddy, had brutally rejected him. For the first time since leaving rehab, he found himself staring at a couple of Vicodin pills in his right hand, with intent. Suddenly, Doctor Cuddy walked into his home and professed her undying love for him. Immediately, his desire for Vicodin evaporated and the pills fell to the floor. The writers incorporated his acknowledgment of the role endorphins played in this improbable occurrence into episode one of season seven. Despite the blooming of their love over a period of weeks, he eventually relapsed, something that she noticed immediately. She had no desire to share him with his other love, so the relationship came to an end, a story that plays out in households around the world with striking regularity.

The Role of Maturity in Addiction

Peele and Brodsky cite Charles Winick, the New York-based Physician and Health Official, who performed an in-depth study of addiction among the users of street drugs. Concerning his research, they state:

> The key to non-addiction is maturity. Winick's discovery that heroin addiction is often an artificial extension of adolescence, an evasion of adult responsibility, offers us a sound insight about addictions of all kinds . . . An essential attribute of maturity is our ability to handle the inevitable conflict between our desire for connection with others and our own individual separateness. . .Your relationship with yourself and your surroundings, rather than any *external confirmation of worth*, is the most secure anchor against perpetual uncertainty and escapism. It is also the

best underpinning for the formation of the relationships which complete an emotionally mature person's life"[4] (my italics.)

The greater part of the book elucidates how we use intimate relationships as an external confirmation of worth (ego inflation by external locus.) What Jesus offered the woman at the well was a relationship with God, a relationship that reconciles our need for connection with a higher power in a way that transcends the limitations of our separateness. These words bear repeating here—John 17:25–26 (NASB):

> O righteous Father, although the world has not known You, yet I have known You; and these have known that You sent Me; and I have made Your name known to them, and will make it known, so that the love with which You loved Me may be in them, and I in them.

Despite our physical separateness, we are never separated from the love of God and if Christ continually dwells in us then, unlike the support provided by other relationships, his support is internal. As the Apostle Paul puts it in Romans 8:38–39 (NASB):

> For I am convinced that neither death, nor life, nor angels, nor principalities, nor things present, nor things to come, nor powers, nor height, nor depth, nor any other created thing, will be able to separate us from the love of God, which is in Christ Jesus our Lord.

The inescapable dependency between love, connection, and the opioids is fulfilled and balanced through an internal relating with the higher power.

Towards the Extinction of Cravings

I have already mentioned the mu opioid antagonist Naloxone, which is effective in preventing overdose deaths, but there is another drug in the same class which is not an effective antidote for acute Opioid poisoning due to its slow onset of activity. *Naltrexone* does, however, reverse both the subjective and objective effects of alcohol by blocking endorphin signaling. Alcoholism is treated using a slow-release formulation of Naltrexone that is injected intra-muscularly once every two weeks. The subject is not required to abstain from drinking during treatment. If they do choose to drink, they will not experience any endorphin-related

---

4. Peele and Brodsky, *Love and Addiction*, 243.

effects and this removes the impetus to drink for many. The goal is to break the mental association between alcohol intake and the feelings of sedation and wellbeing.

Naltrexone's relationship to alcohol is mirrored in Zyban's use in treating nicotine addiction. Zyban (also known as Wellbutrin and Bupropion) is a Nicotinic Acetyl Choline receptor (Nicotine receptor) antagonist which nullifies the limbic response to smoking. Both Zyban and Naltrexone aim to trigger pharmacological extinction events[5], where a subject's normal drug response is snuffed out, leading to a reduction in cravings. If the subject uses a drug for its beneficial mood-related effects, the urge to use will be that much less intense because those effects are impossible to achieve. This approach has proven to be effective for some people.

Abstinence

One salient feature of this strategy is its focus on abstinence as the goal of treatment and this is something that Peele and Brodsky see as unnecessary. They cite research showing that addicts often grow out of their addiction without any counseling or involvement in residential rehab. The clichéd view of the alcoholic is of a person walking on a knife's edge who will lose their balance and fall back into excessive consumption immediately following the first sip of their favorite libation. There are many examples where this is the case, but the research they cite shows this is by no means the majority. They identify cases where, years after recovery, alcoholics can drink socially and in moderation. If we look at relationship as a potential arena for addiction this makes sense. It would not be rational to expect a person who has been involved in an addictive relationship to abstain from relationships for the rest of their life. If the dysfunction in addictive relationships can be remedied, it must also be possible to construct a mental framework that fosters a functional relationship with alcohol. A crucial pre-requisite to this change of perspectives is the desire to change and an open exploration of the pros and cons of such an endeavor (a cost-benefit analysis.)

---

5. Gass and Chandler, "The Plasticity of Extinction."

Addicted to God?

There is, however, one area where Peel and Brodsky's thesis diverges drastically from that presented in this book. They see the invocation of a higher power in the recovery of addiction as the substitution of one dependency for another. This hypothesis may have been born out in their own practice, but on further examination it appears to contradict their assertion that an individual may have either an addictive or a functional orientation in any relationship. In this light it is unreasonable to conclude that every believer is an addict.

An alternative perspective on maturity is presented by Christ and the Twelve Step Program. This view sees maturity as an acknowledgement that life is too complex and unpredictable for even the most capable person to manage and that the only life-serving way to achieve a stable internal state is to acknowledge this. The Twelve Step Program axiom "My life is not within my control" presents a much better starting point than "I am independent and in control my life." The latter statement will ultimately prove false for everyone. At the same time, we cannot claim that an out-of-control life is sustainable either. The involvement of an overarching support, or higher power, is required to give internal structure and to prevent self-structure-disintegration. This may be a relationship with God, a wife, family, friends or community, but nevertheless a relationship.

Wine, Women and Song

The late 1960's and early 1970's saw a Christian revival known as The Jesus Movement. This was a direct counter-reaction to the culture of drug obsession and free love that emerged shortly prior to that time. After experiencing first-hand the emptiness of a hedonistic existence, many determined to fill the hole in their lives with God. The stars of Christian show business (the new wave of contemporary Christian music) often claimed this as their motivation for repentance. Many quit drugs cold-turkey and found fulfillment in the sense of meaning and community that Christ provided.[6]

Peele and Brodsky refer to a conversation with a Christian convert who acknowledged that his experience of relationship with Christ, for the most part, paralleled his addictive relationship with drugs. The

---

6. Race, "A Brief History."

similarities included the emotional highs he experienced during meditation and songful worship. Their conclusion was that religious attachment impedes the individual's ability to create, on their own terms, the self-structures necessary for recovery.

In my view they made a mistake in claiming parity between drug addiction and religious devotion. Their perspective is based on their experience counseling clients for whom the higher power approach of AA had been unsuccessful or even destructive (as was the case for Byron Wood, referenced in our introduction.) No treatment modality will work for 100 percent of patients, so this is not surprising. As with most experiences in life, our perception is colored by the perspective from which we view them. It is true that religious devotion can be a manifestation of addiction, it can stem from a helpless dependency on the external, but it does not have to. The terrain Peel and Brodsky map with respect to intimate relationships contains the same pitfalls, peaks, and troughs shared by any relationship, including a relationship with God. We can relate to God as juveniles or as adults, but we can never deny our physical and emotional vulnerability to the vicissitudes of the natural order. We are never wholly self-sufficient.

The Opium of the People

Karl Marx subscribed to a similar ideology to that proposed by Peele and Brodsky. He believed that religion anesthetizes the pain of the impoverished, allowing them to endure their subjection to the ruling class through their addiction to the faith. The only way he saw to excite the proletariat to rebel against their oppressors was through the repudiation of religion. His goal was to suppress the positive effects of religion on the nucleus accumbens to activate the amygdala, the guard-dog protector and warrior. He needed to make people hurt so much that they had to do something about it, through a process of radicalization. In the introduction to his work *A Contribution to the Critique of Hegel's Philosophy of Right*, Marx penned what has become one of his most oft-quoted passages:

> Religious suffering is, at one and the same time, the expression of real suffering and a protest against real suffering. Religion is the sigh of the oppressed creature, the heart of a heartless world, and the soul of soulless conditions. It is the opium of the people.
>
> The abolition of religion as the illusory happiness of the people is the demand for their real happiness. To call on them to

> give up their illusions about their condition is to call on them to give up a condition that requires illusions. The criticism of religion is, therefore, in embryo, the criticism of that vale of tears of which religion is the halo.[7]

His language here is arcane, but he is saying in essence—religion brings with it a fake happiness like that of the opium high and true happiness can only be found in rejecting the fake happiness of religion. This assertion stemmed from his belief that there is no God, and any such concept exists only in the mind of the believer. Marx also promoted the repudiation of private property ownership. It would seem, on the surface, that this was a rejection of materialism but the passage above reveals his worldview to be one of materialism and naïve realism. In describing the essence of religious thought as faith in illusion he puts *his faith* firmly in the observable and rejects the spiritual realm. He stumbles into the same trap that befalls all materialists in claiming that that which can be observed is the only true reality.

But what of the pursuit of his "real happiness," how successful has that been? And what of those countries that followed Marx in rejecting religion—did they find real happiness and liberty from oppression? Within seventy years of Marx's death 50 percent of the world's population lived in countries ruled by nominally Marxist governments, yet despite Marxism being the ideology of the people, the USSR, China, and Cambodia were the most oppressive regimes of their time, and their inhabitants among the most miserable. On further inspection we see that the happiness promoted by Marx is founded on an external rather than an internal reality and is as fleeting as the joy that flows from a needle in the arm. It is dependent on building a social order that requires constant maintenance, support, and defense from malign natural and geopolitical influences (hard work, like the daily trip to the well.)

At the molecular level, through the internal manifestation of relationship with God, religion *is* the opium of the people, but rather than being sourced externally the opioids that mediate a believer's relationship with their higher power are generated within. The happiness that flows from religious faith is as real as Marx's "real happiness" and both are as immaterial (spiritual) as his "illusion." Both stem from the activity of endorphins. The distinction between them is that a happiness built on materialism (including the material of exogenous drugs) involves constant replenishment from without and is subject to breaks in supply. When the believer's thirst (their psychic vacuum) requires sating, they have at their disposal

---

7. Marx, "A Contribution."

an unlimited supply of living water within that transcends the grave. The water of which Christ spoke to the woman at the well is spiritual, immaterial, and something we can take with us beyond the grave.

Thematic Takeaways

Needs: Our relationship with God meets our safety and security needs in powerful ways. In so doing, God suppresses the amygdala and frees us from the feelings of fear and anxiety it imposes.

Neural Activities: Naltrexone blocks the mu opioid receptor. In this way it prevents the endorphins released through the consumption of alcohol from having any subjective effect. Zyban blocks the nicotine receptor, nullifying the positive effects the smoker seeks to obtain from smoking. The goal of this approach is to bring about pharmacological extinction events, where the individual no longer craves the substance due to its inability to produce rewarding states. This discourages the use of alcohol and nicotine as proxies for relationship and challenges the individual to find mature solutions to their relational deficits.

Relationship: The early stages of an intimate relationship are characterized by states of wellbeing created by the release of endogenous compounds. The body's equivalents of the amphetamines and opioids are the primary drivers of this process. In this early stage of relationship, abstinence from alcohol, nicotine, and other drugs is more easily achieved, but as time passes and relational novelty subsides, the individual may use substances to recapture some of the early magic of love. The endogenous compounds mentioned above may also promote an addictive approach to relationships, where the early excitement of love is sought repeatedly for the high it produces. As we mature, we learn how to sustain high dopamine levels, using more evolved strategies (see self-actualization in chapter 17.) When we are filled with the limitless relational water that God provides, we free ourselves from reliance on external sources of comfort and create a secure attachment with the one who made us.

What you can do now: If your relational issues stem from abandonment trauma, you may benefit from a somatic approach to healing. Look online for free and affordable tools that introduce you to Somatic Experiencing (SE) Therapy, the Hakomi Method, Tension & Trauma Releasing Exercises, and Polyvagal-Informed Therapy.

Part Three—*Integration*: **Developing Sustainable Recovery Practices**

14

Stress

> "Therefore do not worry about tomorrow, for tomorrow will worry about itself. Each day has enough trouble of its own."
> —Matthew 6:34

IN THE PREVIOUS CHAPTERS we examined the cognitive, phenomenological, behavioral, and neurophysiological underpinnings of addiction. Addictions are sparked by our relational dysfunctions and past traumas, but there is a third major factor that predisposes us to addiction—stress. In the remaining four chapters we will consider what can be done to move beyond the destructive coping strategies we adopt in response to these three determinants, and how we can live more contented and productive lives. While I have tried my best to make this last section a profitable read, it can never be more than the porch from which your journey begins, or perhaps a bridge over an obstacle that blocks your path. It is through the investment of energy and commitment that is fully within your control that you will succeed.

The approach presented here builds on our understanding that addictions are strategies that control the balance of activity between the nucleus accumbens and the amygdala. While this model fails to account for the almost infinite complexity of the brain, it is one that I believe fosters an accessible and productive discussion. Addictions provide a means of escape from the unpleasant feelings arising from amygdala activity, most commonly anxiety, worry, sadness, shame, and anger. Factors that chronically stimulate the amygdala might include poor life conditions, trauma and our perceptions of self-insufficiency. It is also our responsibility to distance ourselves from avoidable situations that introduce unavoidable stressors. As John Dupuy points out, for those of us who are predisposed to addiction:

> Stress. Let me state that again. Stress. Chronic, Inescapable, unavoidable stress is public enemy number one in the brain of those who possess the genetic potential for addiction.[1]

While the amygdala's primary role is to inform us of unmet needs through the expression of aversive states, this process becomes dysregulated in the addict. The traumatic emotional memories encoded within it create a baseline state of being that defies time. Rather than healing from our past hurts, they assume the qualities of chronic pain. These injuries were resolved long ago, but the fear of recurrence and the recollection of the suffering these grievances aroused cleave to our identities with magnetic force. Like a statically charged shred of plastic that refuses to separate from our hand, despite our vigorous flicking, these feelings and memories become long-term tenants of our conscious and unconscious minds. The addiction lies in the fruitless process of repetitive flicking and assuaging our unmet needs, in this situation, provides only temporary relief. The primary goal of the recovering addict is to demagnetize the shred of plastic.

A Band-Aid Approach to Managing Emotional Pain

Rather than resolving the chronic emotional pain, the addict leans towards one of the following solutions and often both:

1. Increase nucleus accumbens activity
2. Reduce amygdala activity

In case one, the amygdala remains highly active, but the unpleasant feelings it generates are temporarily masked by the good feelings that flow from the nucleus accumbens. This is the process at work in the abuser of psychostimulant drugs. Solution two lowers amygdala activity to a point below that of the nucleus accumbens, allowing the good feelings the latter provides to predominate. This is the benefit sought by those who abuse sedating drugs like the barbiturates and the benzodiazepines. Alcohol

---

1. Dupuy, *Integral Recovery*, 24.

performs both functions, allowing its partakers to *drown their sorrows* and overcome their social anxieties. It raises nucleus accumbens activity by releasing endorphins and calms the amygdala by reducing neuronal excitability. This is accomplished by triggering the inhibitory action of the GABA receptor, reducing the excitatory action of the NMDA receptor and by curbing voltage-gated calcium ion channel activity. VGCC's are important components involved in neurotransmitter release and inhibiting their action lowers overall electrical activity in the brain. A more life-serving approach, one that avoids the use of exogenous drugs, is to address the issues that are stimulating the amygdala or develop those parts of our brain that regulate its impulsive influence. This will be our goal in this last portion of the book.

The Hormonal Basis of Stress

Cortisol and adrenaline (epinephrine) are portrayed in the media as the fight-or-flight hormones, but often only one of them is framed as such. In reality, they are both involved in the response to threatening situations. They both contribute meaningfully to meeting safety needs in the short-term and their actions are usually self-limiting. Once the threat has been resolved, the levels of these hormones naturally return to normal. However, in our fast-paced twenty first century lives, the constant barrage of responsibilities, activities, and deadlines is perceived by our brains as a continual threat. This can lead to persistently high systemic levels of cortisol which can harm various bodily systems.

Cortisol activates the amygdala, and depending on our genetic and epigenetic profile this might be how its effects are primarily expressed. This kind of stress (distress) manifests in aversive states where fear and worry are the primary outputs. Surprisingly, cortisol also stimulates the nucleus accumbens, leading to a rewarding form of stress (eustress) that motivates our engagement in meaningful activities. Cortisol plays this role in generating the early morning wakefulness that moves us to begin our day. In both roles it acts as a psychostimulant drug, increasing neural activity.

The normal cortisol cycle begins with high levels in the morning that decrease during the day.[2] This is the experience of the *morning person* who feels happy and energetic after rising but low-energy and sleepy

2. Fries et al., "The Cortisol Awakening Response."

in the evening. In contrast, the cortisol levels of the *night owl* increase later in the day and getting out of bed in the morning can be a challenge, when levels are low. It is normal for cortisol levels to rise slightly around three am, but for the insomniac this brings with it a persistent state of wakefulness. During this time the mind spins on a hamster wheel of thoughts leading to one of the radio interviewer's favorite questions "what keeps you up at night"? It is the cortisol that keeps us up at night, although the content of the thoughts can contribute to the persistence of our mental state. This state can be positive or negative depending on our life conditions and how we are wired.

Am I An Anxious Person?

Not everyone is aware that they have an anxiety problem or that it is anxiety that drives their addictive behaviors. Anxiety is not at the root of all addictions, but the evidence establishes it as a prominent component of the addiction process. Admitting that we are anxious people, to ourselves or others, can be a humiliating experience. It took me decades to realize that the feelings I quashed using alcohol were anxiety-related, a realization I resisted with vigor. Anxiety is a close bedfellow of worry, worry being a dysphoric contemplation of *future* possibilities while anxiety is more generalized in nature. Let us begin by considering some strategies for managing worry.

Living for Today

The Bible presents two perspectives on what could be described as "living for today" Matthew 6 exhorts us to put aside future worries and concern ourselves with the present. Live for today, do not be troubled with future matters because there is a God watching over you that has your best interests in mind and remember, if you seek the Divine Kingdom (which is the real reason you are here) there is an eternity ahead of you in which all your needs will be met.

Paul outlines the *other* perspective in 1 Corinthians 15:32 (NIV):

> If I fought wild beasts in Ephesus with no more than human hopes, what have I gained? If the dead are not raised, "Let us eat and drink, for tomorrow we die."

This lifestyle follows naturally from the predicate "the dead are not raised," a condition that reframes life's purpose in profound ways. If the dead are not raised, our reward lies in the pleasure of satisfying our appetites in the present moment. The logical course of action is to indulge ourselves and make the most of today while we still have it. Live as though there is no tomorrow, because the present is all there is. We are free to engage in these behaviors, but freedom carries its consequences.

The True Nature of Freedom

The song *Hard To Exist* by the Spin Doctors (lyrics available online) illustrates the "let us eat and drink" mentality. The first verse addresses boredom, pain, and the vulnerability of our limited physicality, a situation the writer describes as a kind of imprisonment. The second verse provides a potential solution—relaxing on the beach with a pretty girl and a jug of wine. This remedies the situation using external influences; it is a way to bust out of the emotional jail we are in. We could escape using any type of coping mechanism that eases our pain and creates the all-important inner glow. However, the rather unintuitive solution to the conundrum is not busting out, but *busting further in*. You may ask, where is the freedom in that? If I move further in am I not in the sub-dungeon rather than the dungeon? The answer depends entirely on your perspective and the type of freedom you seek.

During one of my failed attempts at smoking cessation I saw abstinence as a huge burden. But after I caved in and began smoking again, I experienced a profound sense of freedom. It is true, I was now free to indulge my cravings, but I was, once again, a slave to nicotine. This paradoxical juxtaposing of freedom and servitude exists in many of life's stations like marriage, work, parenthood, and community involvement. When viewed from one perspective we might perceive a constrained state with more negatives than positives, but by changing the angle of illumination we see the benefits, empowerment, and opportunity our situation provides.

The Prison Inside Me

The most extreme example of this perspective-taking approach I have seen involves the "Prison Inside Me" holiday hotel in Hongcheon South

Korea, where guests *pay* to be placed in solitary confinement for seven days. Their food is passed through a slot in the door, they sleep on the hardwood floor, and they have all the time in the world to unwind and reflect on whatever they need to process. A CBC article[3] about the facility recounts the experience of Suk-won Kang, fifty-seven, an automotive engineer, who described the experience as offering something the outside world lacks—freedom. The operator of the hotel added "Locking themselves up in solitary confinement here is not a prison; the true prison is the world outside." What the hotel's guests are free from, during their stay, is stress, overwork, and the cumulative effects of burnout.

Paul addresses the dual principles of freedom in the context of our faith relationship with God in Romans 6:16-18 (NIV):

> Don't you know that when you offer yourselves to someone as obedient slaves, you are slaves of the one you obey—whether you are slaves to sin, which leads to death, or to obedience, which leads to righteousness? But thanks be to God that, though you used to be slaves to sin, you have come to obey from your heart the pattern of teaching that has now claimed your allegiance. You have been set free from sin and have become slaves to righteousness.

You are always a slave, Paul claims, but you are free to choose your master. Ultimately, freedom vests in the choices we make. Busting out of the prison inside gives us the freedom to gratify ourselves and become slaves of self-indulgence; but busting further in and exploring the intellectual, emotional, and felt-sense space of relationship with the one who is "in us" is true freedom, if we choose that perspective. There is also freedom in sharing this inner life with our communities. God did not intend for us to be hermits, but to shine on the world the light he has instilled in us. As Christians, we can also find inner freedom in borrowed practices from other spiritual traditions, such as mindfulness meditation. Through limited engagement in these practices, we can avoid doctrines that clash with the systems of faith to which we are committed.

Managing Anxiety with Drugs

The treatment of stress and anxiety provided by Western medicine usually involves the prescription of sedating drugs. In the 1950's the Barbiturates

---

3. Kwong, "In a Land of."

were commonly used for this purpose, but the proximity of their therapeutic and lethal doses led to the deaths of many, including Marilyn Monroe and Judy Garland. The barbiturates were later substituted with the benzodiazepines which are better tolerated acutely, but their long-term use is discouraged due to the risk of physical dependency.

We see a similar situation with the opioids which are effective in short-term pain management but often become ineffective or problematic in the long term. These drug families may create the coveted inner-glow, largely due to their anxiolytic effects, but these effects are by no means predictable. For some the opioids contribute nothing but nausea and vomiting, while for others they lift the mood to dizzying highs. Speaking for myself, when given one hundred micrograms of Fentanyl intravenously during a diagnostic procedure I experienced neither nausea nor elevated mood. Bearing in mind the unpredictability of these drug effects and in many cases their short-lived benefits, they fail to provide the freedom from stress and control over mood that we assign to them. There must be a safer and more effective way of allaying our anxieties that avoids the biochemical pitfalls of anxiolytic drugs. Meditation provides such a solution, but while the effects of drugs are almost instantaneous, meditation acts in its role as a long-term lifestyle choice. Its positive effects stem from its ability to promote neural growth in the parts of the brain that express compassion, self-control, and calm.

Living for Today

The idea of digging deeper into our being to fortify ourselves against what is beyond our control is a major feature of many religions. In *The Power of Now*, the spiritual teacher Eckhart Tolle presents a comprehensive system of thought that helps the practitioner rise above the stress and anxiety of everyday life, and reach transcendent states. The title indicates a connection with the admonition of Christ to not worry about tomorrow, but Tolle's approach is to focus only on the present moment. *The Power of Now* promotes a more grounded and less reactive state of being, reactivity being a function of the amygdala that often inflicts damage on relationships. One psychiatrist in our city asks his patients to read *The Power of Now* at the commencement of their treatment program, cautioning them that if they cannot relate to the book's message then he cannot help them. That is certainly some high praise for Tolle's approach, but a little dispiriting to hear at your first appointment.

Breath Work

A central component of Tolle's practice is conscious breath control. This is a common thread seen in the practices of most Eastern religions and some forms of Christianity. There is clear scientific evidence[4] that taking deep breaths and exhaling slowly relaxes the body and reduces stress. A team of Israeli scientists utilized the principle in the development of a breath-control device[5] that provides drug-free regulation of high blood pressure, illustrating the impact this practice can have on our bodies.

Mindfulness Meditation

An umbrella term for strategies like those advanced by Tolle is *mindfulness*, defined by Merriam Webster as:

> The practice of maintaining a nonjudgmental state of heightened or complete awareness of one's thoughts, emotions, or experiences on a moment-to-moment basis

This concept is found in the practices of Hinduism, Buddhism, Taoism, and Yoga and has been progressively adopted as a stress management technique in secular circles. My experience with mindfulness followed a break-in at my home during which thousands of dollars' worth of electrical and computer equipment were stolen. The psychological effects of space-violation caused by a break-in can be profound. Some people address this by re-decorating their homes while others must move to a new location. I stayed put.

Shortly after the break-in, I started to exhibit twitches, or ticks, in various parts of my body. This was a little concerning, but a conversation with my doctor put my mind at rest. She shared a story that drew on her experience in medical school. In classes, she and a friend observed these ticks develop in their fellow students as the semester progressed. It usually peaked with visible symptoms in about 10 percent of the class. She was even able to mimic the different types of tick which I recognized from my own experience. She referred me to a six-week mindfulness meditation course run by the local medical school. Apparently, the elevated stress levels of medical students had been noted, and education in mindfulness had become part of the med school curriculum.

---

4. Zaccaro et al., "How Breath-Control Can."
5. Rosenthal, "Device-guided breathing."

After practicing daily guided meditations for a few weeks I told one of the staff at the gym I attended that it appeared to have cured my lower back pain. Someone overheard me and chimed in "it cured mine too." My ticks also eventually resolved.

Chi Gong Meditation

Another practice that relies on conscious breath control is *Chi Gong*, a modality in which I have some training. This has its roots in Taoism (a philosophical tradition of Chinese origin) and is itself the foundation for Tai Chi. As its name suggests Chi Gong (energy work) manipulates the flow of chi (energy) throughout the body. Although there is little scientific research confirming the reality or nature of chi, I can speak from personal experience that it feels like bioelectricity.

The body's electrical field is the phenomenon measured by the electrocardiogram and electro encephalogram. The same marvel is employed by sharks to locate their prey and by electric eels as a means of self-defense. In contrast to this measurable energetic principle, chi is described as *subtle* energy. At this point the scientific community lacks the technology to explore this phenomenon using the upper-right quadrant approach of objective measurement, but a phenomenological analysis indicates that chi, like bioelectricity, flows through the body's nervous system. Its primary means of locomotion is via the body's central meridian, using the vagus nerves as its main thoroughfare.[6] Chi Gong allows the practitioner to increase the levels of chi at various nerve nexus points, or Dan Tiens, by exercising focused attention at those locations.

The flow of chi is also directed, in Chi Gong, using slow controlled movements. My association of chi with bioelectricity is based on my own felt-sense experience. In the context of stress, however, the electric currents induced by our neurotransmitter system work in tandem with chi at the very least.

The Electroencephalogram

The link between movement and bioelectricity is clearly seen in measurements obtained using the electroencephalogram. These devices include a set of electrodes (thirty-two or sixty-four are commonly used)

---

6. Earth Balance, "Parasympathetic System."

that are placed on the scalp to measure the combined electrical currents of neuron groups beneath the skull. When performing EEG experiments, the desired result is a clean scan. To provide such a result, the subject must keep any kind of muscle use to a minimum. Head swaying, jaw clenching, and any other kind of facial movement will produce an electrical blip on the scan. This blip represents the current that flows from the brain to the facial muscles to trigger a muscle contraction. The most common cause of artifacts in EEG scans comes from blinking, so closed-eye scans are preferred over their alternates. The conscious manipulation of these currents, and their subtle analogs, is the domain of the Chi Gong practitioner.

Energy Cycling

In Chi Gong meditation, mental focus is placed on various dantien energy points. These correspond with ganglionic nerve bundles, for example, the location of the middle dantien corresponds with the celiac plexus, a couple of inches below the sternum. The Chi Gong meditation practice I am most familiar with is energy cycling. A practitioner's energy can be concentrated in the lower dantien (three inches below the naval) using deep breathing and by focusing awareness on that locale. From there it can be moved around the perineum to the lower back, up the spinal column, over the top of the head, and back through the middle dantien to the lower dantien. If performed repeatedly this will cause a release of tension in various muscle groups and a generalized reduction in stress. My Chi Gong coach was able, through this process, to cure himself of chronic asthma.

The energy cycling practitioner must prevail over the three blocks (chi blockages.) The first block is in the pelvic floor, the second in the anus and the third in the lower back. They are three areas where stress-induced muscle tension commonly occurs. These are located between the lower dantien and the mid spine.

In Chi Gong terminology, muscle tension in these areas is caused by an overabundance of chi, triggered by a block in flow. In neuro-physiological terms stress-related lower back tension is caused by electrical signals from the autonomic nervous system firing endplate-potentials in the muscles surrounding the lower spine:

End plate potentials (EPPs) are the depolarizations of skeletal muscle fibers caused by neurotransmitters binding to the postsynaptic membrane in the neuromuscular junction. They are called "end plates" because the postsynaptic terminals of muscle fibers have a large, saucer-like appearance.[7]

Warning—A Too-Much-Information Moment Approaching

I write this section, despite its unseemly subject matter, as an example of how meditation can ease stress-related health issues. Muscle tension in the urinary tract and alimentary canal is the result of overactivity in smooth muscle fibers. There is no comparable structure to the end-plate in smooth muscles—receptors for neurotransmitters are found throughout the cell membrane. The importance of managing stress-related muscle tension is highlighted in that back pain is the second most common cause of workplace absence. Back pain may be due to a physical injury (a pinched nerve or herniated disk perhaps) but in many cases there is no detectable physical cause.

Meditation can provide a long-term solution in these cases, without the need for drug interventions. The Western medical approach may be used to reduce levels of chi in the pelvic floor by disrupting autonomic signals using a Muscarinic receptor antagonist such as Fesoterodine. Hemorrhoids, a common cause of discomfort and distress, present a more nuanced case. At the risk of a too-much-information moment I'll briefly explain their most common cause: muscle tension in the smooth muscles of the anus prevents blood from flowing through the veins from outside the body to the inside, causing it to back up and create swellings. This condition can be treated with a nitric-oxide donor such as nitroglycerine which, through membrane interactions, initiates a signaling cascade within the muscle cell. This breaks the myosin light-chain, causing the muscle to relax. This is why nitroglycerine is effective in reducing the symptoms of angina, as it relaxes the smooth muscles of cardiac blood vessels.

Using nitroglycerine as a treatment for hemorrhoids was encouraged by the Harvard educated Doctor Gabriel Mirkin on his daily radio show, but it is not a treatment accepted by some doctors. Nitroglycerine relaxes all smooth muscles in the body, including those in the blood vessels of the brain. The amount of nitroglycerine administered should

---

7. Wikipedia Contributors, "End-plate potential."

be kept to a minimum to avoid headaches and, in cases of overdose, cerebral edema.[8] Meditation provides a drug-free solution to these muscle-tension related maladies, the goal being to move damaging levels of chi from tense muscles to one of the dantiens or by cycling it through the central meridian. The ability to do this can be achieved with just a few weeks of daily meditation practice.

Feldenkrais

Feldenkrais is a modality that treats the musculoskeletal dysfunctions that arise from the long-term activation of our fight-and-flight response. If our safety needs are not met by our life situation then our bodies will be in a constantly guarded mode. Our muscles will be bracing against the perception of imminent threat. Feldenkrais teaches the patient to feel this process in action, to be aware of it, and address it before the negative health effects of muscle tension manifest themselves. One practitioner I know put it this way:

> Our bones are quite capable of passively supporting the weight of our bodies, but we learn, through the anticipation of threat, that this is the job of our muscles

These fear-related strategies are rooted in the self-protection instincts of the amygdala. What we see in the meditation practices of Eastern religions are strategies for managing this unruly influence, which is something that all religions contribute to the world.

Non-reactivity and non-violence

Jesus made non-reactivity (non-violence) a foundation principle of his teaching on the Kingdom of God in Matthew 5:43–46 (NIV):

> You have heard that it was said, 'Love your neighbor and hate your enemy.' But I tell you, love your enemies and pray for those who persecute you, that you may be children of your Father in heaven. He causes his sun to rise on the evil and the good, and sends rain on the righteous and the unrighteous. If you love those who love you, what reward will you get? Are not even the tax collectors doing that?

---

8. Carl, "Intravenous Infusion."

It is disappointing that many Christians see *reactivity* as part of their religious observance, in becoming the guard dogs for the community, always resisting people, behaviors or ideas they see as problematic. In its macroscopic form this tendency has caused unbelievers to label religious movements as violent by nature. Religious violence is not caused by any belief system but rather by the pack instinct of the amygdala which is fed and nurtured through a process of radicalization.[9] As Andrew Newberg and Mark Walkman observe in their book *How God Changes Your Brain*:

> The problem isn't religion. The problem is authoritarianism, coupled with the desire to angrily impose one's idealistic beliefs on others.[10]

They go on to discuss their research into religion's role in a unique neural circuit with involvement from the PFC and other structures, that:

> Enhances social awareness and empathy, while subduing destructive feelings and emotions.[11]

The influence of meditation on the development of this circuit is a central theme of their book. Rather than inhibiting the amygdala by stimulating the nucleus accumbens, meditation loosens the influence of both over the thinking and actions of the individual by promoting neuroplastic growth in the pre-frontal region (the Human) and another brain region that lies sandwiched between the PFC and the limbic system, the Anterior Cingulate Cortex.

The Anterior Cingulate Cortex (ACC)

The ACC is one of the primary structures that mediates between the PFC and the limbic system. It is also particularly vulnerable to the ravages of drugs of abuse. Damage to the ACC creates long-term disinhibition, breaking the lines of communication that allow the Human to exercise influence over the Chimp. It also contains the Brodman Area 25 (or subcallosal cingulate), the target of Selective Serotonin Reuptake Inhibitor drugs.

I have not, up to this point, devoted much time to a discussion of serotonin and its role in mediating reward or its role in addiction. While

---

9. Canna and St. Clair, "Neuroscience Insights."
10. Newberg and Waldman, *How God Changes Your*, 11.
11. Newberg and Waldman, *How God Changes Your*, 14.

serotonin is active in attraction and the establishment and maintenance of relationships, it does not play much of a role in addiction. As UCSF Neuro-endocrinologist Robert Lustig explains, dopamine and serotonin implement separate reward systems. The mantra of dopamine is "this feels good, and I want more" while for serotonin it is "this feels good and I don't want more"[12] High serotonin levels create a state of satiation rather than one of yearning, so the effect of serotonin-inducing drugs is to dissuade their acute use rather than to promote it.

In addition to the generation of rewarding feelings, an active Brodman Area 25 does share one other function with the nucleus accumbens—it inhibits the amygdala.[13] Meditation, through the promotion of neural growth in the PFC and ACC, diminishes the feelings the addict seeks to escape. This has the potential to permanently resolve the effects that traumatic memories impose on the individual and stop the hamster wheel of negative thoughts about the self and others. Meditating on the word of God is an extremely powerful catalyst for neural growth in the ACC, as is Christ's practice of loving those whose goals and beliefs do not align with our own. Conversely, engaging in aggressive thoughts, words and actions towards other people and situations enhances amygdala function.

Selective Serotonin Reuptake Inhibitors

A quick functional boost, which can form a stop-gap strategy while neural growth in the ACC takes place through meditation, is the administration of an anti-depressant drug. My own experience with Prozac is a case in point. When commencing a course of treatment, something I did a few times, my anxiety was reduced (my amygdala inhibited) by more than half within a few hours of taking the first pill. This observation goes against the standard medical wisdom that SSRI drugs take weeks or months to work. While this might be true for their role in depression, their effects on anxiety appear to be almost immediate for some. This is the principle behind their off-label use for treating premenstrual stress and premenstrual dysphoric disorder.[14]

---

12. Lustig, *The Hacking of*, 8.
13. Newberg and Waldman, *How God Changes Your*, 17.
14. Freeman, "Premenstrual Dysphoric Disorder."

As with most drugs we have discussed, SSRIs can also have paradoxical effects for some people, so proceed with caution. Serotonin is involved in the release of adrenocorticotropic hormone which acts with the hypothalamic-pituitary-adrenal axis to increase cortisol levels. For this reason, SSRIs will increase anxiety for some when first administered, but this effect usually diminishes over time. My experience illustrates even more unpredictability. Raising serotonin signaling levels using SSRI drugs reduces my anxiety levels, while raising them by ingesting serotonin's metabolic precursor 5-hydroxytryptophan has the opposite effect. While 5HTP is used by some as a sleep aid (it is also the metabolic precursor of melatonin) it promotes agitation and insomnia for me. Life truly is a box of chocolates—you never know what you are going to get.

I want to emphasize again that while the amygdala plays a role in our many dysfunctions, it is not evil by nature; it is there to alert our higher functions to the presence of a threat, which is a necessary survival tactic. The trick to living in agreement with Christ's admonition to love our enemies is to use our rational processing abilities to connect reactivity with reality. Jesus's counsel in Matthew 5 may also require action that hurts the actor (integrity,) noting that repayment will be made for such sacrifices in the form of an eternal reward.

Meditation in the Bible

While meditation is mentioned approximately twenty times in the English Bible (depending on the translation) it is always in the context of meditating on God's words. There is no emptying of the mind as there is in Eastern meditation. Meditation is not explicitly observed by most Christian denominations and when it is it might diverge from the picture painted in the Bible. Laurence Freeman compares Christian meditation with mindfulness as follows:

> In mindfulness techniques the attention remains on yourself (thoughts, feelings, sensations, body scan.) In meditation the work is to take the attention off yourself. This is the simplest and hardest thing in the world to do and yet also the most transformative and liberating.[15]

He also describes meditation as "a prayer of the heart." The Quakers are another historically Christian group that emphasizes the importance of

15. Di Pietro, "Sensing God."

meditation in worship. To them it can be an emptying of the mind, primarily in order to listen for God to speak, but can also be a process of reflecting on how God has worked in their lives.

Meditation aside, the main anxiolytic effect of faith in Christ comes from doctrine rather than observance. Eastern religions take a "how" approach to attaining a more healthy and grounded mental state, while Christianity takes more of a "why" approach:

> Why worry when your father in heaven will provide (Luke 12:22-26.)
>
> Why fear punishment when Christ died to cover your sins (Romans 8:1.)
>
> Why be anxious? Cast all your anxiety on him because he cares for you (1 Peter 5:7.)
>
> Why fear death when Christ offers you everlasting life (Mattthew 19:29.)
>
> Why fear loss when what you lose you will find (Matthew 10:39.)

The advantage is gained not so much in doing as it is in trusting.

So why do we return to our old ways of coping even after giving assent to our reliance on a higher power? In NVC terms this is referred to as strategy attachment, and it is a central feature of the addiction experience. Recent research has shown the guard dog analogy of the amygdala to be applicable in this context too. When it bites, it doesn't want to let go:

> The hippocampus and the amygdala store information about environmental cues associated with the desired substance, so that it can be located again. These memories help create a conditioned response—intense craving—whenever the person encounters those environmental cues.[16]

We can substitute the phrase "desired substance" with "desired behavior," showing the applicability of this attachment process to non-substance-based addictions like sex and gambling.

As described in the article, the anticipation of the reward experienced by the nucleus accumbens, combines with the intensity and fear the amygdala expresses against abstinence to create a persistent obstacle to recovery. Only when sufficient supports are in place can the reliance

---

16. Holtzhausen, "Addiction."

on substance and behavioral strategies be overcome. In the Christian context this could be described as letting go and letting God—it is all about surrender.

The Importance of a Good Night's Sleep

In chapter 11 we discussed the role of sleeplessness in promoting emotional lability. Sleep is an undeniable human need, and when that need goes unmet the amygdala becomes agitated. Sleep refreshes our minds and bodies in many ways. One of the most important to our mental health is the clearing of glutamate, lactate, amyloid-beta, and other neuropeptides (metabolic waste) from our cerebrospinal fluid.[17] The dysphoria that stems from sleepless nights could be treated with psychostimulant drugs, but this compounds the problem by making sleep more difficult to attain. It is not uncommon for the psychostimulant addict to go for a week without sleep. Good sleep hygiene is a principle that aids in recovery, firstly by removing the need to take sedatives to sleep and stimulants to wake up, but also through soothing the amygdala by giving it what it needs to calm down. You cannot have a stress-free day if you have accumulated high levels of sleep debt.

My advice, from personal experience is to avoid sleeping pills unless you use them for less than two weeks at a time. I briefly used benzodiazepines and hypnotic agents as sleep aids, but the hangover the day after was always problematic. For this reason, I have followed a meditative approach to sleep management. Using these medications to take a two-week break from sleeplessness to catch up on your accumulated sleep debt might pay off on occasion.

The Sleep Stages

A good night's sleep unfolds in various stages (NREM means non-rapid-eye-movement.)

1. NREM Stage 1 (N1): Lightest stage of sleep.
2. NREM Stage 2 (N2): A deeper stage than N1 but still relatively light.
3. REM Sleep (R): Often considered lighter than NREM Stage 3 but deeper than N1 and sometimes N2 in terms of the ease of waking.

17. Chong et al., "Sleep, Cerebrospinal Fluid."

4. NREM Stage 3 (N3): The deepest stage of sleep, also known as slow-wave sleep (SWS.)

While the analysis of sleep stage data was the purview of the sleep therapist until recently, advances in wearable technology have placed consumer grade EEG devices within reach of the public. One example of such a device is the Muse S headband, a product of the Toronto-based company Interaxon.[18] While I have, to this point in the book, been unable to conduct my own upper-right quadrant analyses of neural function, the Muse S empowers me to perform objective measurements of my subjective states.

The Muse S Headband

The Muse S was designed to be worn comfortably around the head during the night. While lab-grade EEG devices have dozens of electrodes, it has only four—a Frontal Lobe and Temporal lobe contact on each side of the head. Despite its simplicity it has been used in a research setting by the Mayo Clinic, Harvard University, and NASA, among others.

The app that accompanies the Muse S implements two main algorithms—a calm algorithm for capturing and analyzing meditation sessions and a sleep-stage algorithm. The latter can identify:

1. Wakefulness
2. Rapid Eye Movement sleep
3. Light sleep (N1 and N2)
4. Deep sleep (N3)

Figure 14.1 shows a recent sleep session I recorded that helps illustrate the role stress plays during the night.

---

18. Ghosh et al., "Mindfulness Using."

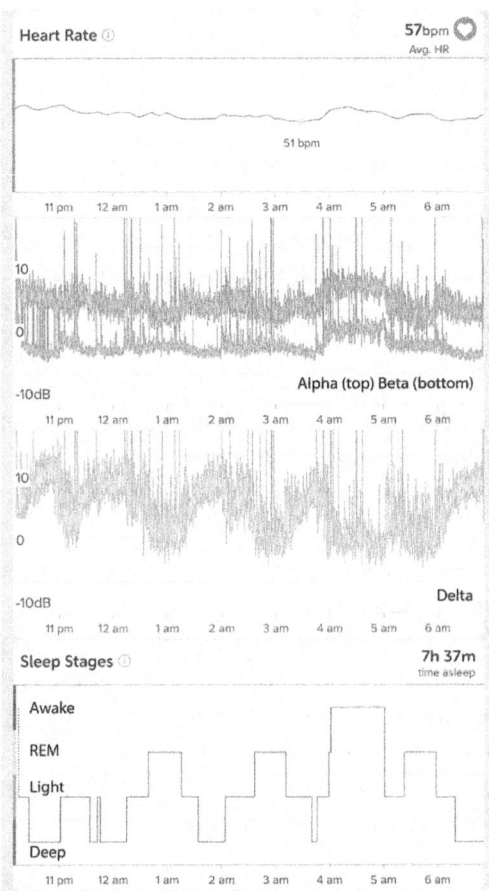

Figure 14.1 A Personal Sleep Stage Session. EEG Data Recorded Using the Muse Headband. Image © Interaxon Inc., used with permission.

The Muse app interprets sleep states from the raw EEG data. I have included three brain-wave frequency bands that provide some insights, along with some basic heart rate data. Generally speaking, high levels of low frequency waves correspond with deeper sleep states (slow wave sleep,) while high levels of faster electrical oscillations indicate a wakeful state. The frequency ranges included in the figure are:

1. Delta waves: 0.5 to 4 Hz (sleep)
2. Alpha waves: 8 to 13 Hz (calm wakefulness)
3. Beta waves: 13 to 30 Hz (active thinking and focus)

Examining the data from a phenomenological perspective we see that I experienced a period of wakefulness between four am and five am (triggered by a rise in cortisol.) This is reflected in a rise in the alpha and beta frequencies and a drop in the intensity of delta waves. (A correlation is visible in the delta wave and sleep stage graphs between low levels of delta waves and wakefulness.) At the same time there was a moderate rise in my heart rate. This was accompanied by an intense feeling of eustress as I anticipated the tasks that awaited me in the day ahead. On days when this state is particularly intense, it resembles the feeling I had as a kid early on Christmas morning! While I sometimes lie awake at night for an hour or two, I am never subject to feelings of distress during those times, for which I am thankful. I still experience distress at times during waking hours, in response to the normal challenges of life, but not at night.

While cortisol is one component of the flight or fight response, the fact that my heart rate rose only moderately, during wakefulness, indicates that my adrenaline levels remained low. As discussed in chapter 2, adrenaline has a profound effect on heart rate and cardiac stroke depth. The intensity of eustress experienced during this session highlights the subjective effects of cortisol on the nucleus accumbens.[19] As a result, I find it difficult to let go of the rewarding phenomena expressed during these wakeful periods, making my insomnia a kind of addiction. I know I could use mindfulness as a sedative, but I do not do it because that would mean letting go of those rewarding feelings.

Hypnagogic Sleep

The current iteration of Interaxon's sleep-stage algorithm is not very proficient at distinguishing N1 from N2 type sleep, so both are classified in the sleep stage graph as light sleep. For most insomniacs, their condition is exacerbated by the fear that they will not be able to get back to sleep. This becomes a self-fulfilling prophecy (an example of paradoxical intention) as the rising cortisol levels engendered by this fear prevent their transition into slumber. However, for the past couple of weeks, my sleep graphs have shown no wakeful periods at all, even though I experienced periods of consciousness during the night. This indicates that my subjective perceptions of conscious awareness were during periods of N1 sleep, I was only half awake.

19. Stephens and Wand, "Stress and the HPA."

N1 sleep (also known as Hypnagogic sleep) includes periods of hypnagogic imagery—vivid sensory experiences, such as visual images, sounds, or feelings, that occur as one is falling asleep or transitioning between sleep and wakefulness. These can sometimes feel like daydreams or hallucinations and may be a mix of reality and imagination. The consolation here is that periods of N1 sleep are still restorative to the mind and body. This knowledge alone can prevent the fear reaction that arises when the insomniac becomes conscious of their surroundings in the early hours of the morning. An acceptance that consciousness does not in itself indicate another impending sleepless night may be enough to prevent the transition from N1 into wakefulness. Even when fully awake, lying still with closed eyes is still restorative. Insomnia is only as damaging as we perceive it to be. Accepting nighttime periods of conscious awareness as normal puts us on the path to better overall sleep performance.

Cortisol and Addiction

Having discovered the link between cortisol and my nighttime periods of euphoria, my mind turned to a consideration of cortisol's potential synergistic effect when combined with psychostimulant drugs. I had noticed that when engaging in stressful work situations, the influence of my ADD medication, Ritalin, on nucleus accumbens function was greatly amplified. In one case I had to restrict my use of the medication to four pills a week rather than two per day. The subjective effects were extremely distracting, and I did not want to use a perfectly valid treatment to escape the pressures of the project. Research has shown that in addition to amplifying the limbic effects of drugs like cocaine, high cortisol levels can trigger a relapse in a recovering user.[20] Animal studies have shown that when serum cortisol concentrations meet a certain threshold, the animal will begin to self-administer cocaine, while in the absence of cortisol this activity was absent.[21]

I had also observed that the limbic effects of Ritalin were, for me, most noticeable in the winter, peaking in the month of February. This is likely because cortisol levels follow a seasonal pattern due to winter changes in temperature, freedom of movement outdoors and daytime

---

20. Graf et al., "Corticosterone Acts."
21. Goeders, "Stress and Cocaine Addiction."

light levels.[22] This is particularly true for those who, like me, suffer from seasonal affective disorder.

The core message is that when life conditions are most stressful, we are at the greatest risk for relapse. We do not just turn to drugs to drown our sorrows; it is when we are most stressed out and miserable that drugs have the effect we are looking for. For this reason, an active approach to stress reduction prior to a period of abstinence is of great importance.

Seeking Calm Through Christian Meditation

After using the Muse headband for tracking mindfulness meditation sessions for several months, my mind turned to a consideration of Bible-based meditation practice. Bearing in mind the subjective benefits I accrued from my love affirmation (described in chapter 12) and the release of endogenous opioids it engendered, I decided to track a session using the Muse app's calmness algorithm. Figure 14.2 shows the results.

I conducted a ten-minute session consisting of five minutes of free-form thinking followed by five minutes of repeating my love affirmation and expressing thankfulness to God for his gifts, guidance, protection, and oversight. It took a minute or so for the warm wave of relationship-affirming opioids to commence, but after that my mind remained consistently in the calm zone. This was a mind-filling exercise as described by Laurence Freeman above, rather than a mind-emptying one, but the calming effects were the same as those I experience using a mindfulness approach. Once in the calm zone, the app provided auditory feedback through a tropical bird soundtrack, as indicated by the blue bird icons below the graph.

This exercise provides an objective measure of the stress-relieving benefits of meditation. When stress levels drop, the amygdala is inhibited, cortisol levels are diminished, and the conditions are established for neural growth in the Anterior Cingulate Cortex and PFC. Over a period of days or weeks of meditation practice, this will strengthen the inhibitory influence of ACC and PFC over the limbic mind and increase subjective perceptions of peace and contentment. I highly recommend that all those who rely on substances or behaviors for support establish a meditation practice prior to embarking on their recovery journey.

---

22. Sher, "Cortisol and Seasonal Changes."

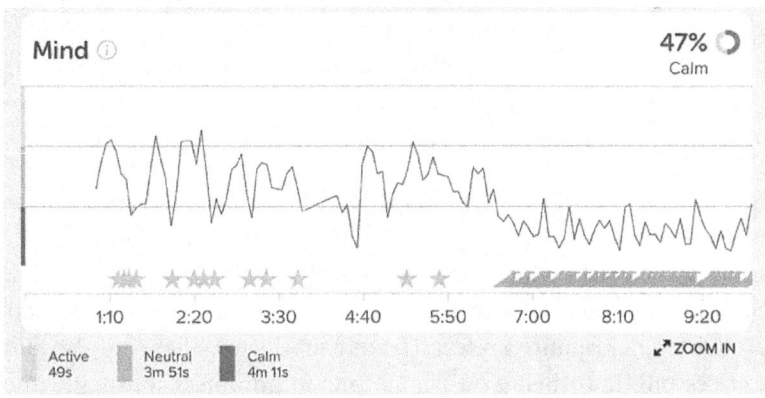

Figure 14.2 The Calming Effects of My "Love" Affirmation. EEG data recorded using the Muse headband. Image © Interaxon Inc., used with permission.

Thematic Takeaways

Feelings: Addictions are strategies that control the balance of activity between the nucleus accumbens and the amygdala. Activity in the nucleus accumbens generates good feelings while activity in the amygdala promotes feelings of fear, anxiety, anger, sadness, and shame. These are the feelings that drive the compulsive behaviors that characterize addiction.

Needs: Freedom is a primary need that finds expression through autonomy, choice, ease, and independence. At the core of freedom is the ability to choose what we want for ourselves. The choices we make can be quick fixes or long-term strategies. They can also be destructive or life-serving in nature. Solutions with an internal locus (enjoying the prison inside) provide the greatest benefit in terms of autonomy.

Neural Activities: Cortisol stimulates both the amygdala and the nucleus accumbens giving it the dual roles of motivational coach and bodyguard. It moves us to action with feelings of engaged anticipation but also notifies us of potential threats. The Brodman Area #25 is part of the anterior cingulate cortex where neural growth is accelerated by mindfulness meditation. It is also the target of SSRI antidepressants and a powerful inhibitor of the amygdala. Promoting neural plasticity in BA25, through meditation, reduces stress and promotes feelings of contentment, addressing many of the feelings that drive addiction.

Relationship: Meeting our relational needs (especially the one we have with our higher power) reduces stress through the release of endorphins and other relationship affirming compounds. A strong belief that we are lovable and loved promotes a state of calm and loosens the grip our destructive strategies have on us.

What you can do now: If you want to connect with others in person, check your local listings for free Chi Gong practice sessions. These are commonly offered by local organizations in public spaces. Mindfulness meditation classes may also be offered for free in your area, although these sometimes require a referral from your doctor. There are also ample resources online to help you begin your mindfulness journey without leaving the house. Several companies manufacture EEG devices that help track your meditation and sleep performance. These are not an affordable option for all, but some manufacturers offer payment plans which can help. These devices provide a whole new perspective on the mind that can help to motivate you and accelerate your healing.

15

An Integral View

"When we understand that each and every aspect of our humanity is imperative to feel whole, we take the evolutionary leap into self-love and living our most authentic life."
—Debbie Ford

IN COGNITIVE TERMS HUMANS tend to compartmentalize. We identify ideas and perspectives and classify them as compatible or incompatible with our self-concept and system of values. Often the distinction is made on moral grounds. We tag the ideas of God, family, purity, joy, and peace with positive attributes while denying the influence of greed, sensuality, violence, and anger in our lives. We are averse to situations where these competing influences come into contact. We put one set of influences in a box labelled "good" and another in a box flagged "bad" and hope they never touch. The juxtaposition of opposites such as peace/violence and purity/sensuality repel us, and when they do interact, we experience shame.

If we are supposed to embody the attributes of the "good" box and shun those of the "bad," how can we be comfortable embracing the reality of both expressed in our lives? An honest assessment of the situation would convince us that we do, in fact, embody both sets of influences, but for the sake of our own self-perception, and to project an image of respectability to others, we hide the contents of our "bad box" in the furthest recesses of our mind where it can wreak havoc. These influences are manifestations of what psychologist Carl Jung termed the *Shadow*—the embodiment of our dark side.[1]

These unconscious shadow aspects of our psyche contribute to feelings of self-loathing; they reduce the good feelings that flow from a healthy self-esteem and promote the unpleasant feelings that emerge when esteem needs are not met, all without our conscious awareness. As a result, we might attempt to generate good feelings by engaging in addictive behaviors. The approach to addiction recovery covered in this chapter is known as shadow integration and involves accepting the unpresentable characteristics that we hide from ourselves and others.

The Shadow

A good starting point for discussion of the shadow is the contrast between the propensities of the Limbic System and Prefrontal Cortex. There is more to the shadow than this, but please bear with me while we consider another pop-culture analogy. Season one episode five of Star Trek the Original Series (*The Enemy Within*) illustrates the PFC/limbic distinction quite well. Kirk (captain of the Starship *Enterprise*) is beamed up from the surface of a planet and arrives on the Enterprise as two different physical individuals. Their arrival is offset by a few seconds, so nobody is in the transporter room when Kirk #2 arrives, and he is able to blend into the normal buzz of ships operations unnoticed. This situation does not last long, however, as Kirk #2 tries to assault a female crew member and receives some nasty scratches to his face. Meanwhile Mr. Spock (second officer) has noticed that Kirk #1 is acting strangely. He seems listless, unmotivated and is unable to remember basic operating protocols. Eventually the dichotomy between the two Kirks becomes common knowledge among the ship's crew and, after some analysis, Dr. McCoy identifies the following attributes possessed by the two:

Kirk #1

- Love
- Compassion
- Intelligence
- Self-control

1. Ford, *The Dark Side*, 24.

Kirk #2

- Aggression
- Decisiveness
- Passion
- Guile

These characteristics are the basic attributes of the Pre-Frontal Cortex (Human) and Limbic System (Chimp) respectively. Kirk #1 was incapable of performing his leadership role as Captain, while Kirk #2 roamed the ship's halls like a wild animal assaulting security details and officers alike, including Kirk #1. The solution, proposed by Spock and Scotty (Chief Engineer,) was to reverse the polarity of the transporter and put both Kirks back through in the hope that the two personalities would be merged back into one individual. This (spoiler alert) strategy turns out to be successful and an *integrated* version of the captain walks off the transporter platform, with confidence and grace, to resume his directorial duties.

On the surface the moral of the story is that the more bestial aspects of our nature are as necessary to our functioning as are the human ones and that the only way a person can function is by integrating both aspects into their being. Having been raised by parents who played the role of our PFC until ours was fully formed and integrated into our being (a process of emotional maturation) we learned to have a very poor view of those behaviors, beliefs, and attitudes that were so frequently censured in our youth. One focus of this parental nurturing process is the frequent admonition to care for the feelings of others. This exhortation towards empathy lays the foundation for many of the skills necessary to function as members of a social group, be it a family, collegial troop, religious community, gardening club, or society as a whole, and yet we all fail repeatedly. We cheat, violate social mores, lie, let our loved ones down and feel guilty about it all.

Resistance is Futile

This sense of guilt causes us to resist the more beggarly qualities we see in ourselves, but rather than addressing their manifest reality we resist their expression *in others*. This external-focused aversion creates a destructive cycle of tension that both exacerbates our personal situation and reduces our ability to redress it. Jung encapsulates this truth in the statement:

> What you resist not only persists but grows in size.

Herein lies a key point—the use of the word "resist" differs from its use in James 4:7 (NIV) where James states:

> Resist the devil, and he will flee from you

Jung is not saying that the only way to keep our limbic system in check is to stop resisting it. The resistance to which he refers could better be described as denial. Rather than accepting the truth that we share the propensities and capabilities of the rapist, murderer, and thief, we deny (resist) this reality. Because we hide these tendencies behind a mask which turns both inward and outward to disguise our true nature from ourselves and others, we become blind to it. But this does not prevent us from seeing *our* shortcomings in others. We become judgmental and condemn the people in our lives while being blind to those same principles at work in us. As Jung put it:

> . . .it is quite within the bounds of possibility for a man to recognize the relative evil of his nature, but it is a rare and shattering experience for him to gaze into the face of absolute evil.[2]

This quality of self-observation (shadow perception,) rarely observed in humankind, is a prerequisite to the acceptance of reality. Rather than familiarizing ourselves with the contours and features of the face described by Jung (our true moral appearance) our tendency is to put it out of our minds, and do as described in James 1:22–24 (NIV):

> Do not merely listen to the word, and so deceive yourselves. Do what it says. Anyone who listens to the word but does not do what it says is like someone who looks at his face in a mirror and, after looking at himself, goes away and immediately forgets what he looks like.

As bizarre as it sounds, this level of self-deception is pretty much the default mode of human operation. We all tend to see ourselves as decent upstanding citizens, while it is those around us that need rehabilitation. Jesus speaks of this axiom in Matthew 7:3–5 (NIV):

---

2. Jung, *Collected Works*.

> Why do you look at the speck of sawdust in your brother's eye and pay no attention to the plank in your own eye? How can you say to your brother, 'Let me take the speck out of your eye,' when all the time there is a plank in your own eye? You hypocrite, first take the plank out of your own eye, and then you will see clearly to remove the speck from your brother's eye.

We cannot see the plank in our own eye because we deny (resist) its existence. It has become part of our unconscious rather than our conscious awareness. Perhaps this is the quintessential application of the word *hypocrite* (play actor.) We are playing out the role of the righteous one in our own minds when our flaws are at least equal to those of the people we disdain. That plank in our eye used to be a speck, but our denial of its existence allows it to grow without restraint. Ironically, we cannot address the faults we refuse to acknowledge.

Jung's Imago

What Christ is conveying through this metaphor is Jung's concept of the imago.[3] This is an image of another that we project onto them, unaware that it exists within our own unconscious. This imago may have no relation to the person we accuse, rather it describes aspects of our own personality that we fail to recognize. We might apply this process to someone with whom we are in conflict, accusing them of perfidious thoughts and actions, when those transgressions exist only within us. In these situations, our accusations give others a transparent view into our unconscious, in which the denunciations we throw at the other are actually admissions of our own guilt. This is clearly illustrated by Paul in Romans 2:1 (NIV):

> You, therefore, have no excuse, you who pass judgment on someone else, for at whatever point you judge another, you are condemning yourself, because you who pass judgment do the same things.

This is a first century expression of the modern idiom "every accusation is an admission." When our negative judgements of others prove to be inaccurate it can be extremely embarrassing, should we have the humility to notice. If we take the time to assess the origin of our condemnation, we are likely to find the same transgression within ourselves. The power of the

---

3. Ulanov and Ulanov, *Religion and the Unconscious*, 224.

imago in self-development is in allowing such realizations to reframe our perception of our own standing in relation to those around us.

The Mind of the Flesh and the Mind of the Spirit

A major component of Christ's ministry was the delivery of this message to the people, and he used the Jewish leaders of the time as his prime examples. Their challenge in recognizing their own plank was twofold: firstly, they were entrusted with interpreting a law which defined the parameters of righteous living. This enabled them to mold each precept to their own advantage and ensure that their Chimp remained satisfied while at the same time they were abiding by the rules in their own minds. Secondly, they believed it was possible to perfectly observe their interpretation of this law and that their job was to be seen by their followers as doing just that. It is hardly surprising that Christ describes them as hypocrites and blind leaders of the blind. The ability to deny their own shadow was practically part of their job description. Nowhere is this presented more clearly than in the parable recounted in Luke 18:9-14 (NIV):

> To some who were confident of their own righteousness and looked down on everyone else, Jesus told this parable: "Two men went up to the temple to pray, one a Pharisee and the other a tax collector. The Pharisee stood by himself and prayed: 'God, I thank you that I am not like other people—robbers, evildoers, adulterers—or even like this tax collector. I fast twice a week and give a tenth of all I get.' "But the tax collector stood at a distance. He would not even look up to heaven, but beat his breast and said, 'God, have mercy on me, a sinner.' "I tell you that this man, rather than the other, went home justified before God. For all those who exalt themselves will be humbled, and those who humble themselves will be exalted."

Denial is the wool the Chimp pulls over the eyes of the Human. It allows the Chimp to operate in stealth mode. But where in the Bible do we find a description of this Limbic/PFC dichotomy? The Apostle Paul describes the two warring minds as the mind of the flesh and the mind of the spirit in Romans 7:5-9 (NIV):

> Those who live according to the flesh have their minds set on what the flesh desires; but those who live in accordance with the Spirit have their minds set on what the Spirit desires. The mind governed by the flesh is death, but the mind governed by

the Spirit is life and peace. The mind governed by the flesh is hostile to God; it does not submit to God's law, nor can it do so. Those who are in the realm of the flesh cannot please God. You, however, are not in the realm of the flesh but are in the realm of the Spirit, if indeed the Spirit of God lives in you.

This passage provides some context for the title of this book—The Science and Spirituality of Addiction. It frames the struggle between the Human and Chimp, as a spiritual conflict. All our addictive behaviors are rooted in the mind of the flesh, a mind that operates independently of the desire for self-improvement and self-preservation. The mind of the flesh is death in both the temporal and the eternal sense. Even in an illegal drug market, such as that in North America—which has been contaminated with lethal levels of Fentanyl—addicts continue to use, knowing that their next fix could be fatal. Feeding the appetites of the limbic loop has become, for them, more important than life itself. The strength of a Christ-centered approach to recovery, as promoted in the Twelve Step Program of Alcoholics Anonymous, is that it brings the spiritual nature of the conflict into full view, thus promoting life and peace.

Jung's Views on Spirituality

You might wonder what a panspiritualist like Jung has to offer to this spiritual view of our inner struggles. While Jung might be ignored or criticized by many Christians, my assertion is that this is largely a function of their own shadow. It is true that Jung was disillusioned with Christianity, but this is not surprising as what he saw in his father and six uncles (all Swiss Reformed pastors) was an overemphasis on externals and intellectual beliefs at the expense of a true heart-connection with God and the people in their lives.[4] This is the folly of over reliance on the lower-right quadrant of Wilber's AQAL matrix, an obsession with rules and regulations while ignoring the subjective experience of the individual and how this manifest in the lower-left quadrant of community. This is the very quality that Jesus struggled to address in the religious leaders of his time, and on this point Jung and Christ agree—legalism does not equal spirituality. This is a failing of organized religions of all stripes, and a major factor behind western culture's drift towards a relational/decentralized church rather than one with a

---

4. Wehr, *Jung: A Biography*.

centralized/bureaucratic structure. In the anthology C. G. Jung Letters Volume I[5] Jung states his position as follows:

> The main interest of my work is not concerned with the treatment of neurosis, but instead with an approach to the numinous [Transcendent God experience]. The approach to the numinous is the real therapy, and inasmuch as you attain to numinous experience, you are released from the curse of pathology. Even the very disease takes on a numinous character!

What he asserts here is that not only is a transcendent spiritual experience key to overcoming our mental dysfunctions, but the dysfunctions themselves exist in a spiritual space. This is as true of our addictions as it is of our neuroses, depression and anxiety.

The higher-power approach to overcoming our compulsions and addictions need not be Christ-centered. Alcoholics Anonymous gives the practitioner freedom in choosing the nature of their higher power. It need not be framed in religious terms, but it must be something to which faith and commitment can be applied. The higher power can be a cause, an institution (such as marriage,) a belief (perhaps political in nature) an activity or long-term project.

An Alternative Higher-Power Approach to Recovery

A few years ago, I attended a nine-day International Intensive Training (IIT) in non-violent communication where I met a woman who had been addicted to crystal meth for years and who had funded her habit by engaging in prostitution. Towards the end of this period, she became involved with the Aikido community and its practices. Aikido is a modern Japanese martial art developed by Morihei Ueshiba as a synthesis of his martial studies, philosophy, and religious beliefs. It soon became clear to her that crystal meth addiction was incompatible with Aikedo practice, and because Aikido met her need for mind-body connection and provided a framework for spiritual expression she was able to let go of her destructive coping mechanisms.

The key contribution afforded by a focal influence such as this is stability. A bicycle is inherently unstable because its weight rests on two points of contact with the ground. Balance can only be maintained while in motion due to the gyroscopic effect of the spinning wheels and because

---

5. Jung, *Letters of*.

the back-wheel trails behind the steering axis. Likewise, if the only rubber hitting the road of life vests in the limbic/PFC dyad, the tendency will be towards collapse. If, however, these are attached to a third point of contact, the system becomes self-supporting. Using the Chimp Paradox system, devised by Dr. Steven Peters, the third point of contact vests in the memory. This is where the conscious embracing of life-lessons, beliefs, values, and commitments coalesce into the autopilots. Engaging with these autopilots is what Andy Stanley encourages in his affirmation:

> I will prioritize what I value most over what I want now.[6]

For a higher-power approach to be effective we need to create an attachment to our guiding principle of choice, but if we remain subservient to the Chimp, we will never grasp this opportunity. Submission to our craving for autonomy and control over our own destiny limits our strength to the resources we possess as an individual. Surrendering control to a guiding ethic and letting *it* lead expands our capabilities enormously, but this cannot take place without external help. Peters devotes an entire chapter, *The Troop Moon*[7], in *The Chimp Paradox* to the establishment of relationships with those people who possess the resources we need to achieve success. The paradox here is that external influences are required to develop an internal locus of control. Regular meetings with an individual's fellow addicts and mentors and the ability to hear and be heard fulfill this role in the context of the Twelve Step Program.

An Approach to Shadow Integration

Prior to her death from cancer in 2013, at age fifty-seven, author Debbie Ford explored the shadow and the higher-power principle in detail. In her book *The Dark Side of the Light Chasers* she introduces a system that helps an individual confront their shadow, gaze into its eyes, and accept it as an indivisible and vital component of their being. Her book describes the methodology she used in her own life to achieve reconciliation with her shadow and how she took these discoveries on the road, helping many others to achieve internal harmony through attending her workshops. Despite her lack of formal education in the field, her works have been incorporated into teaching materials of universities and colleges.

---

6. Stanley, "How to Get."
7. Peters, *The Chimp Paradox*, 154.

One thing she alludes to repeatedly is her own experience with drug addiction and how this was addressed through engaging in shadow practice. In her twenties Ford was addicted to cocaine and the opioids, sometimes consuming as many as one hundred pills per day. She attended rehab on several different occasions before she was able to rein in her compulsive behaviors.[8] During the rehab process she learned much from the stories of others during group counseling sessions, and these unique insights into the lives of addicts gave force to her therapeutic method.

The Tension Between the Shadow and the Light

Ford theorizes that many of our emotional, relational, and spiritual problems arise from the tension that exists between our light and shadow aspects. As shown in our Star Trek example, the separation of Kirk's warring factions into two physical bodies not only accentuates the operation of Limbic system and PFC as separate minds but also highlights the enmity between the two. Kirk #1 disdains the crass and aggressive behavior of Kirk #2, while Kirk #2 mocks the passivity and weakness shown by Kirk #1. Yet to deny the existence, indeed the pre-eminence, of the limbic mind in our lives does nothing to reshape the neuronal infrastructure on which our consciousness is built. As the prophet writes in Jeremiah 13:23 (NIV):

> Can an Ethiopian change his skin or a leopard its spots? Neither can you do good who are accustomed to doing evil.

In the same way as skin and hair color are genetically predetermined, we are genetically hardwired for moral struggle. To be more precise, part of our brain bestows important instinctual drives which, if left unchecked, bring about destructive outcomes and another brain region grants the ability to regulate those drives. The regulated party sees the other as an unnecessary and confounding influence on its self-serving agenda and the regulator sees the other as the cause of all its problems.

This physiological approach gives us a physical/functional perspective on the shadow. In contrast, Ford takes a more semantic approach where words reveal the realities behind our inward-facing mask through the emotions they arouse in us. In her workshops Ford presented a list

---

8. Yardley, "Debbie Ford."

of over 250 words that are common triggers, and asked her clients to identify which of these generate the most noticeable emotional charge. Here are a few that I found to be particularly triggering:

Liar

Phony

Hateful

Vindictive

Wimp

The Advantages Bestowed on Us by Our Weaknesses

My mask causes me to deny (resist) the features in myself described above and convinces me that I am a generous, honest, loving, easygoing, resilient, and forgiving guy. Shadow practice, however, sees every person as a manifestation of a universal hologram. If you were to cut the hologram on your credit card into tiny pieces, each piece (representing an individual) would contain the entire image. In the same way, every person contains the entire gamut of human capabilities and propensities. We are all capable of exhibiting the shadow attributes listed above, but few of us would willingly embrace our shadow and in so doing "stare into the face of absolute evil." The key to accepting these attributes in ourselves comes from recognizing and accepting the gift that each one bestows. The first point in the list corresponds with the ninth commandment "thou shalt not lie," and yet the Bible provides many examples of the benefits of lying.

Joshua 2:4–6 (NIV) recounts a situation where two spies were sent to assess the city of Jericho and its defenses. They entered the house of Rahab and evaded capture through the lie she told:

> But the woman had taken the two men and hidden them. She said, "Yes, the men came to me, but I did not know where they had come from. At dusk, when it was time to close the city gate, they left. I don't know which way they went. Go after them quickly. You may catch up with them." (But she had taken them up to the roof and hidden them under the stalks of flax she had laid out on the roof.)

This lie paved the way for the overthrow of the first city to the West of the Jordan. 2 Samuel 17:20 recounts the story of Jonathan and Ahimaaz, who evaded capture by King Saul's men when Bahurim's wife hid them in a well and scattered grain over the cover. When asked where the men were, she replied: "They crossed over the brook."

Had they been caught, King David would not have received important intelligence about an impending attack by his son and rival, Absalom. Moving on to the next word "phony" we can see that David gained much by *pretending* to be insane. 1 Samuel 21:10–13 (NIV) recounts the following:

> That day David fled from Saul and went to Achish king of Gath. But the servants of Achish said to him, "Isn't this David, the king of the land? Isn't he the one they sing about in their dances: "'Saul has slain his thousands, and David his tens of thousands'?" David took these words to heart and was very much afraid of Achish king of Gath. So he pretended to be insane in their presence; and while he was in their hands he acted like a madman, making marks on the doors of the gate and letting saliva run down his beard.

Being hateful towards our fellow man can cause much conflict and injustice, but God is hateful towards the haughty eye, hands that shed innocent blood and those that devise wicked schemes (Proverbs 6.) Being vindictive over small wrongs committed against us may eat away at us every waking moment, but as revealed by L. Sun in his book *The Fairness Instinct* the idea of vengeance plays a huge role in almost every culture[9], indeed Exodus 21:23–25 (NIV) requires it:

> But if there is serious injury, you are to take life for life, eye for eye, tooth for tooth, hand for hand, foot for foot, burn for burn, wound for wound, bruise for bruise.

The law also prescribed the establishment of six "cities of refuge" to which a person who had killed another accidentally could flee to avoid the avenger of blood. In cultures where vengeance is observed in this way it becomes a self-perpetuating system of justice, freeing the government from playing the role of arbiter in most cases.

The Benefits of Being a Wimp

The examples above show how the words *liar, phony, hateful,* and *vindictive*, all of which created a shadow-oriented emotional charge in me (because I deny their reality in myself) each have a gift to offer when the

---

9. Sun, *The Fairness Instinct*, 131.

right situation arises. While looking to the Bible as an inspired source of guidance on the positive aspects of the shadow, the word *wimp* does not appear in any Bible search. For that reason, I must look to my own life experience, and those of others, to see what gifts that concept bestows on the lives of every person. An attendee, Steven, at one of Debbie Ford's workshops struggled with this exact concept, being unable to see how being a wimp (something he had denied in himself for his whole life) could possibly contribute anything. After thinking for a while, however, he realized that it had probably saved his life. When he was in college he was out drinking with friends for a couple of hours when one of them suggested they drive to a bar in a different town. Because of their recent alcohol consumption, he was uncomfortable with this and decided not to join them. His initial thought about this fear reaction was that it reflected badly on his manliness, but his perspective changed when he learned that the car went off the road a few minutes after they departed. One of his friends was killed and the other three were seriously injured.

In integrating my own inner wimp, I recalled an experience I had when in my mid-teens. While waiting at a bus stop with my girlfriend, a punk rocker jogged past. He had an idiosyncratic running style, so after he passed, I mimicked it to get a laugh. Unbeknownst to me, a friend of his was following and witnessed my display of mockery. This lad had the typical skin-head look—Doc Martin boots, skinny jeans, graphic tee shirt and no hair. He kicked me, expecting to get a rise out of me, but I did not react. He demanded "come on then, start on me" (start a fight) but I responded "no." This confused him, so he asked "why"? I answered, "because I'm a pacifist." He kicked me again and left. While the contest was not exactly David vs. Goliath, I was certainly outmatched and could have been seriously injured had I risen to the challenge and not wimped out.

Am I ugly?

Another word that used to trigger me was "ugly." The charge I associated with that word came largely from an interaction I had with one of the *popular* girls in high school. My family and I attended a performance of the Mikado at the local theatre. To my surprise this seventeen-year-old girl was a member of the cast. The following week I bumped into her in the hallway at school and complemented her on her performance. She smiled kindly and thanked me, but as she walked away, she commented to her friend "did you hear what that *thing* said"?

The memory of this judgement, that I was just a thing, set the bar for my self-assessment for years. What I needed was desensitization to the emotional impact of this memory. This is one function of integrating a trigger word using Ford's approach, but it can also be accomplished using other modalities. I overcame my emotional charge to "ugly" when my wife performed an NVC role-play session with me. She played the role of the girl and answered any questions I had for her about the incident. When I asked her "why would you call me a thing"? she responded:

> I do regret saying that, and I really didn't want to hurt your feelings. You probably don't understand how difficult it is being the popular one. Popularity is a brutal thing and sometimes you must be brutal to maintain your place in the pecking order. If my friend thought that I was talking to you because I was seeking the company of one of the un-popular boys, it would have been a strike against me. I had to distance myself from you to protect my position. I hope you can understand.

When embracing conversations such as this one, it does not matter if the perspective presented in the role-play is true. My wife had no idea whether the girl was thinking along those lines or not, but my choice to believe that this was a possibility made all the difference. The limbic system tends towards black and white thinking. From a shadow viewpoint I might be objectively ugly, from a light perspective I might have a beautiful soul. The truth of the matter inhabits more of a gray area, as they say: "beauty is in the eye of the beholder." Acknowledging this gray area is typical of an integral approach. It disconnects us from our attachment to emotional, semantic, and perceptual absolutes and inures us to other possibilities. From the integral perspective we are both worthless and worthy, ugly and beautiful, lazy and conscientious.

The Integration of Perceived Ugliness

In *The Dark side of the Light Chasers* Ford deals with the same issue in her own life.[10] Because her father addressed her using pet names like pig-nose and bucky, she assumed she was an ugly person. This set her on a journey to prove to *herself* that she was beautiful by engaging in a long string of relationships. The problem with this strategy is that it fails to address the root cause of the belief. No matter how many men she

---

10. Ford, *The Dark Side*, 136.

found that saw her as attractive she still needed another one to prove it. This did not work as she was still attached to an absolute perception of ugliness. What did work for her was engaging in a desensitization exercise to reduce the emotional charge that "ugly" generated in her. She wrote down five potential reasons why her father used those pet names, three of them positive and two negative:

Positive

1. I am beautiful, so my father became nervous around me. The only way he knew how to deal with his nervousness was by calling me names that he thought were cute.
2. My father thought these names were cute and used them with affection.
3. My father loved me so much that he wanted to prepare me for the real world. He thought he could protect me by downplaying my beauty.

Negative

1. My father hated me and was trying to damage me for life.
2. My father thought I truly was remarkably ugly and the only way he could deal with it was by teasing me.

The goal of this exercise was to choose an alternative interpretation to her father's use of the names. She used the following line of reasoning to select positive point three above.

I always ask myself, "Does this interpretation empower me or disempower me? Does this interpretation make me feel weak or strong?" If you have an inner dialogue that disempowers you, it will not change until you yourself replace it with a positive, powerful, internal conversation.

These internal conversations are equivalent to Dr. Steven Peter's autopilot concept. They are knowledge fragments that form the baseline for our interpretation of various life contexts. They also correspond to the Latin word *conscientia* from which the English word conscience is derived. Conscientia refers to a knowing within ourselves. As such our conscience is that mode of thinking that reflects our beliefs and values as opposed to what we desire in the present moment. The conscience draws on that part of our memory that is accessible to both the limbic and pre-frontal regions, and which mediates between our impulsive and regulatory functions. Yet we tend to choose what we want in the moment rather than what we value and suffer the consequences. As Ford states, in relation to our negative self-talk and self-perceptions *"we are addicted to pain and suffering."*

The Joy of Self-Deprecation

Consider this example of the addiction to suffering. After I identified the subjective sensation of dopamine reward, which everyone can do by observing and cross-referencing sensations brought about by rewarding experiences, I realized that I could generate a hit of dopamine through self-deprecation. This usually involved a response to a mistake that I had made or a judgment I had received from a third party. Although the most notable emotions at the time might have been anger, shame, and embarrassment, these were accompanied by a warm dopamine glow. The inner dialogue that fed this limbic release was a belief that I had done what I *shouldn't* have done, justice had been served and that I got what I *deserved*. Rather than resist this dialogue I would double down in judgement on myself. Research Shows[11] that guilt and shame activate the nucleus accumbens (the limbic system's main pleasure center) just as effectively as pride does. In NVC terms, the words "deserve" and "should" are highly effective at throwing our mental constitution off balance, and yet the shame they generate within us can be addictive using the same mechanism as that is employed in cocaine dependence.

Eye Movement Desensitization and Reprocessing (EMDR)

Another modality for desensitizing an individual to the emotional charge provoked by certain words is EMDR which is commonly used for the treatment of Post-Traumatic Stress Disorder.[12] EMDR follows a similar approach to Ford's, but the focus is initially on traumatic events rather than words. Following the selection of an event an associated word, or words, that produce an emotional charge are chosen from a list and focused on while a bilateral form of stimulation is employed. The term EMDR was initially coined because the practitioner asks the patient to follow an object with their eyes as it is moved from left to right

---

11. Dion, "Are Guilt."
12. American Psychological Association, "Eye Movement."

close to their face (picture the cliched image of a hypnotist swinging a pocket watch from side to side.) The practitioner will often use their hand for this purpose, but this exposes them to the risk of developing repetitive strain injury, so vibrating balls that switch on and off alternately in the right and left hand are also used. This may be accompanied by sounds that play alternately in the left and right ears. These activities provide visual, tactile, and aural inputs that alternately cross the body's meridian. This switches the processing of these inputs between the left and right brain hemispheres in quick succession.

While Ford's approach addresses the differences in logic between the mid and forebrains EMDR considers that each brain structure (for example the PFC, nucleus accumbens, amygdala, and hippocampus) has a left and right-side analogue, hence EMDR integrates both fore and mid brains as well as the left and right hemispheres. The eye movement, sensation, and audio exercises, which last a couple of minutes each, are repeated until the chosen words no longer generate an emotional charge.

All too often our addictions are a response to the painful sensations associated with this charge. Once the charge has been dealt with, there is no longer a need to address the pain—it has been resolved. EMDR is another integral approach in that it resolves traumatic memories by welcoming them into the realm of consciousness. Symptoms like rage and flashbacks only persist while these memories are suppressed and pushed into the unconscious.

Nicotine Addiction

Nicotine addiction is a habit with some of the worst health outcomes of any form of drug use. The first time I tried a cigarette I was hooked and as nicotine was the only drug I had ever used, the whole experience was novel and exciting. What surprised me was the similarity between nicotine stimulation and the sensation of being in the arms of a loved one. This was great! (Like the apostle Paul does in 2 Corinthians 11, I speak here as a mad man.) With a cigarette I could get my need for feelings of intimacy met without investing any energy in cultivating a relationship! I was also 100 percent in control of this inanimate partner which had no preferences and made no demands. Nicotine became, for me, a stand-in for authentic relationship.

I observed a similar preoccupation with control in an interaction I had when asking for a light from a middle-aged man at a bus stop. We got into a conversation about our reasons for smoking and he spoke with some passion about how smoking gave him control over his life. Having been raised by an authoritarian father, who despised smoking, the practice allowed him to exercise a measure of autonomy while at the same time "sticking it to the man." When I shared with him my struggles around abstaining from nicotine use, he replied "I will never quit smoking, because the day I do my dad wins!" This is a tragic result of a son's rebellion against the imago the father attempted to impose on him[13] using the traditional worldview's domination mode. His reaction would have been quite different had he seen his father's wise admonition as a safety net (which rescues) rather than a fishing net (which entraps.) Both the father's imago projection and the son's reaction were expressions of their respective shadows. The unconscious coping strategy adopted by the son had immediate emotional benefits but blocked the development of his relationship with the father and himself. (It has also caused the rupture that divides many an atheist from their higher power.) As Carl Jung wrote:

> Until you make the unconscious conscious, it will direct your life, and you will call it fate.

The way in which we both used nicotine has one thing in common—neither approach addressees the root problem. Smoking cannot resolve the feelings of self-consciousness, anxiety, and inadequacy we might experience in our pursuit of a relationship, neither can it heal the relational hurts accumulated during adolescence. What it does do, however, is further distance us from resolution and/or acceptance of our psychological realities. We have now overlaid on our insecurities a significant financial burden and exposed ourselves to serious cardio-vascular, respiratory, and mutagenic hazards. In refusing to see the true nature of our struggles we exacerbate the problem. In *resisting* their reality, through smoking, our ability to address them is curtailed, so they are free to grow unhindered. In contrast, bringing them into consciousness allows us to see them as they are, analyze them and embrace the triggers that brought these insecurities into existence.

---

13. Ulanov and Ulanov, *Religion and the Unconscious*, 224–226.

Thematic Takeaways

Feelings: There is great freedom in accepting that we are all capable of engaging in behaviors we hate and that when we engaged in them in the past it was for beneficial reasons. If we deny the existence of these tendencies in ourselves, they will continue to exert their influence on us from within our unconscious minds. Because we cannot accept these principles in ourselves, we label them as "other" (second or third person) and project them onto our fellow humans. This introduces tension into our lives, makes us reactive and promotes feelings of shame, anger, anxiety, and sadness. These are the feelings we seek to escape through our addictions.

Neural Activities: A good, but incomplete, example of shadow formation involves the clash in values between the prefrontal cortex and the limbic system. The Human (PFC) disdains the law-of-the-jungle practices of the Chimp (limbic system) and assigns them to others rather than acknowledging them as integral components of the self. Yet every human has a limbic system and is capable of those behaviors. Acknowledging the chimp (looking into the face of pure evil, as Jung put it) brings an end to the unconscious warfare between the two and promotes harmony in the mind/body system.

Relationship: Because the rewarding phenomena that build relationships flow from the activities of dopamine, any drug that boosts dopamine function can mimic the subjective experience of relationship. This includes nicotine, alcohol and the amphetamines. A relationship with a higher power opens a doorway through which to explore the numinous (transcendent spiritual) experience of recovery. This relationship helps build the self-structures (autopilots) that give strength to our frame and save us from falling. With these structures in place our addictive strategies lose their value.

What you can do now: Look for free and affordable shadow work apps. These help you to embrace the disowned aspects of yourself that are draining you of energy and creating mental turbulence. This activity corresponds with the "clean up" step of Ken Wilber's four step strategy for self-actualization mentioned earlier in this chapter. You may prefer using an app, but you might also consider engaging in Wilber's *3-2-1 Shadow Work Process* with a pen and paper. You can find a good introduction to this at www.drbren.com/blog/3-2-1-shadow-work-process.

Shadow material often includes unresolved traumatic experiences. EMDR can help process the emotional charge around these memories or traits. For this reason, you may find that using an EMDR app in combination with your shadow practice is beneficial.

16

Growing Up

"The key to nonaddiction is maturity. Winick's discovery that heroin addiction is often an artificial extension of adolescence, an evasion of adult responsibility, offers us a sound insight about addiction of all kinds." —STANTON PEELE: *LOVE AND ADDICTION*.

VIEWING ADDICTION AS MERELY a function of molecular chemistry on the human nervous system is extremely limiting. This perspective exists at the physical level and ignores the psychological and spiritual perspectives. We have explored the physiological/functional divide between mid and forebrains and the role of language as a tool for integrating these attributes into a holistic view of the self. This chapter builds on this concept and introduces a holarchic model which describes the eight known levels of human central nervous system development. Rather than dividing the brain into different functional units which perform discrete tasks this approach explores the changes that transpire from neural growth, cell migration, myelination, synapse formation, and learning. These occur progressively throughout the developing brain from the day we are born until the day we die. It is how we grow up.

The development of the human consciousness unfolds through phases extending from the sensory-motor states of infanthood to a condition of maturity that includes and transcends all previous phases. The mature individual exhibits features from each of these developmental phases and cannot exist in isolation from the synthesis of all preceding stages. The simplicity afforded by the limbic/pre-frontal dichotomy in its role in regulating behavior is part of this, but in the context of addiction we need to move beyond this and consider the phases of developmental psychology. The limbic brain contributes attributes such as pleasure-seeking behavior, aggression, and fear to the human condition, but this is its function in all humans. My goal in this book is to identify why this process goes off the rails for some of us and not others—and what to do about it.

The Holon

Describing these phases of development is simplified using Arthur Koestler's concept of the *holon*.[1] A holon, simply put, is something that is both a whole in and of itself while simultaneously being a part of something else. Clear examples of this idea can be found in the field of molecular physics. An atom is composed of a nucleus and an electron cloud. The nucleus is formed from a balanced distribution of neutrons and protons. The nucleus is a discrete entity, but it is built using protons and neutrons. Each of these, in turn, are formed from quarks. The quark, proton, neutron, nucleus, and electron clouds are components that constitute an atom. The atom is a thing which includes and transcends all these underlying components. If the nucleus ceased to exist so would the atom. If the proton ceased to exist so would the nucleus and the atom. If the quark should disappear so would proton, neutron, nucleus, electron cloud, and atom. Physiologically speaking we could extend this *holarchy* upward through the molecule, cell, and member to form an entire human body. The body could not exist without the concept of a member. It is true that some members can be removed without affecting the body's existence, but if the *concept* of the member ceased to exist the body would disappear.

This holarchy concept can be applied to any ordered system. At each level we find a measure of strength, utility or influence on the resulting whole that increases as the levels progress. Take the rope for example:

---

1. Panarchy.org, "Arthur Koestler."

Figure 16.1 The Rope as a Hierarchy of Holons or Holarchy. Wikimedia Commons.

The rope is a holon with four levels. At the most basic level is the fibre. These are grouped together into treads with a diameter of approximately two millimeters. Threads are twisted around each other to form strands, with a diameter of roughly ten millimeters, and strands are twisted around each other to form the rope. From the strength/influence perspective the fibre has very little intrinsic value. You might be able to lift a few grams of material with it before it breaks. The thread contains many fibers and inherits the combined strength of each one, allowing you to lift several hundred grams. The rope, however, might support an object weighing a ton. The fibre has very little contribution to make to the task of lifting weight, but it has enormous *extrinsic* value. If the concept of the fibre ceased to exist so would the rope. This is an integral view of the rope—each level in its anatomy includes and transcends the previous levels, and so it is with our levels of conscious existence. No matter how mature we are, we still rely on the most basic aspects of our humanity and rejecting those baser attributes brings instability to the mind/body system (which could manifest as shame.) The strength of the individual as a conscious entity depends on the extent to which the previous levels have been integrated into the whole.

Ken Wilber's Levels of Personal Development

In the previous chapter we identified the strength that comes from consciously including these baser attributes as foundation principles for the development of a healthy psyche. In his book *A Brief History of Everything*[2] Ken Wilber describes an eight-level *ladder* of consciousness development that starts with the sensory-physical infant dimension

2. Wilber, *A Brief History*, 127–36.

and progresses to the level of spiritual transcendence. This corresponds with what I have referred to, informally, as the worldview-zeitgeist spectrum. While this is based on the same concept of shadow integration presented in the previous chapter it embodies a perspective with much higher resolution. Using Wilber's model, we are no longer limited to a dark/light (monochrome) view of consciousness, but one that exhibits various hues. But before we dive in, let us explore some of the principles on which Wilber's theories are founded.

Depth Psychology as a Foundation for Personal Development

In the early part of the twentieth century two great minds featured prominently in this discussion—Sigmund Freud and his protégé Carl Jung. These individuals believed that the key to understanding the pathologies of the present lay in the experiences of the past. Some (most notably Czech psychologist Stanislav Grof) drew on Christian tradition in support of their hypotheses. In Matthew 18:2–4 (NIV) Jesus says:

> He called a little child to him, and placed the child among them. And he said: "Truly I tell you, unless you change and become like little children, you will never enter the kingdom of heaven. Therefore, whoever takes the lowly position of this child is the greatest in the kingdom of heaven."

While I take this as an exhortation to acknowledge our complete childlike dependence on God and the humility and surrender that such a perspective affords, to others it indicates that true spirituality and healthy psychological functioning can only be attained by returning to the depths of our nascent existence. Using this approach, personal development includes three components—the perinatal, the personal, and the transpersonal, the transpersonal being the deepest.[3] I find it more relatable to think of these stages as the pre-rational, rational and post-rational. The rational embraces the principles of scientific thought and it is from this perspective that I have written much of this book. Upon fully embracing the rational, however, its limitations become evident, and consciousness progresses to a space that closely resembles the perinatal with its subjective rather than objective attributes. This *depth psychology* approach underpins the work of Freud, Jung and Grof.

---

3. Grof, "Beyond Psychoanalysis."

Grof's model described in *Realms of the Human Unconscious*[4] focusses largely on the perinatal where he believes that memories from shortly before and after birth and especially the birth experience itself, hold the key to healthy spiritual development. From this perspective spirituality takes root before birth and progresses through the personal (rational) before making a U-turn back to the transpersonal. This view shares concepts with Otto Scharmer's U-Theory of management.[5] While returning to the vertex of the U corresponding to the transpersonal (which inhabits the same psychological space as the perinatal) could be construed as a complete regression, the U is often represented as a spiral, the transpersonal vertex being higher than the perinatal, indicating a more mature perspective on the same space. Picture this U-shaped spiral as a ramp between levels of a parking lot which rises from the previous level and curves through 180 degrees, giving the driver access to a higher point over the same two-dimensional space. With the psyche, this higher vantage point allows the individual to view the attributes of consciousness that were implicit shortly before and after birth through the lens of life experience.

While largely adhering to the views of Jung, Wilber identifies in these ideas what he calls the pre-trans fallacy. To Wilber spiritual development is progressive rather than regressive. We move ever upward from the perinatal to the transpersonal. Wilber represents this process as a ladder with nine rungs but admits it could also be portrayed as the progressive layers of an onion, or the segments of a Russian doll where each layer envelops and includes the previous ones. Each rung of the ladder or layer of the onion has its own value system or world view. He describes this as a *height psychology* approach. Wilber also divides the process of personal development into four phases[6]:

Grow Up by moving through the early stages of emotional maturing

Clean Up by doing shadow-up [shadow work]

Wake Up by doing spiritual practice

Show Up by serving humanity in the world

---

4. Grof, *Realms*.
5. Scharmer, *The Essentials*.
6. Zweig, "Ken Wilber's Call."

Spiral Dynamic Theory

One integral technology that draws on, and extends, these ideas is Spiral Dynamic Theory (SDT.) Rather than viewing neural development as a simple U-shaped spiral that begins and ends in the same conceptual space, SDT combines the idea of continuous upward motion expressed in Wilber's ladder with the spiral. In SDT terms we are constantly advancing in our development and understanding of ourselves, the social and cultural environments in which we exist, and how we relate to those environments. Our conscious development consists of a series of curves in the spiral, first progressing to new territory and then curving back to view our old stomping grounds from a higher perspective (see Figure 16.2.) The upward curve into new territory provides the developmental fodder our minds crave. We chew on it and ingest it but are unable to digest it without curving back and integrating it into the self through a process of re-assessment. This is the process of maturation or growing up.

Turquoise – Holistic Self
Collective Individualism - starting 30 years ago
Authentic irony. An ecology of perspectives. Deep experience of oneness.
Quest: restoring joy and thriving to the whole spiral.
Method: deep interior work; working with intuitive knowing, paradox, multarities;
power through presence and radiance; dancing with what arises.
Gifts: experience intuitive and collective intelligence; return of Purple at higher level through ritual, joy, awe, and profound stable connection to all that is.

Yellow – Integral Self
Meshworks, Flex and Flow; Interlocking Dynamic Systems - starting 50 years ago
Head with heart. Repeating beige at higher level of self-sense:
survival of human species. Healthy hierarchies. Integrate the whole spiral.
Quest: spiral health; address toxicity & ensure development at and through each stage.
Method: leading self first to interior clarity then acting with elegance.
Gifts: decrease in fear; flexible perspective-taking; increase in cognitive complexity (e.g. polarity management); increased emotional and spiritual intelligence (EQ and SQ); creative solutions that address multiple levels of the spiral; form fits purpose.

2nd Tier

Green – Sensitive Self
Social Democracies, informational, Pluralism - starting 150 years ago.
Seek inner peace within a caring community. Power with, Solidarity, Human Rights Activism.
Quest: dialogue and consensus; care for all; care for the planet.
Method: appreciate diverse views, listen well, consensus, emphasize group needs, marginalize no one.
Gifts: empathy and social skills (EQ); concern for ecosystems; social justice; tolerance; pluralism.

Orange – Rational Self
Democracy, Science, Capitalism - starting 300 years ago
Act from self-interest by playing the game to win. Market-Driven Meritocracy
Quest: create a better life through logic, reason and strive-drive.
Method: science, learn to excel, set goals, achieve, measure success.
Gifts: individual rights; democracy; modern medicine; industrialization; critical thinking (IQ); early global perspective.

Blue – Rule/Role Self
Monarchies, Nation States, Authoritarian Religion - starting 5,000 years ago
Ethnocentric, life has meaning, direction, and purpose with predetermined outcomes.
Quest: order and purpose; defining good and evil; safe pathway to the good.
Method: fit in, follow the given rules, don't exceed your role, discipline, faith
Gifts: deferring gratification; impulse control; rules and roles.

Red – Power Self
Feudal & Exploitive Empire - starting 10,000 years ago
Aggression, might makes right, be and do what you want, regardless.
Quest: warrior status, power, glory, self-expression.
Method: align with power, take what you need, power over others, force.
Gifts: individuation, passion, creativity, courage, boundaries, risks.

Purple – Magical Self
Tribal Order - starting 50,000 years ago, Egocentric, impulsive
Keep the spirits happy and tribe's nest warm and safe. Magical thinking
Quest: safe mode of living, security.
Method: Petition to Gods or Powers with ritual.
Gifts: awe, wonder, ritual, fun and play.

Beige – Instinctive Self
Survival Bands - starting 250,000 years ago
Do what you must to stay alive. Un-differentiated, narcissism.
Quest: food, water, warmth, shelter, reproduction.
Methods: communication by signals and sounds but not language, scavenge whatever you need.
Gifts: survival instincts; intuitively sensing trouble.

1st Tier

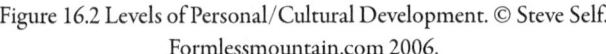

Figure 16.2 Levels of Personal/Cultural Development. © Steve Self. Formlessmountain.com 2006.

The Work of Don Beck, Christopher Cowan and Clare Graves

Don Beck and Christopher Cowan developed the ideas underlying SDT, building on the work of Clare Graves, during the last quarter of the twentieth century. Beck also worked with Ken Wilber to incorporate SDT into the AQAL matrix as figure 16.3 shows.

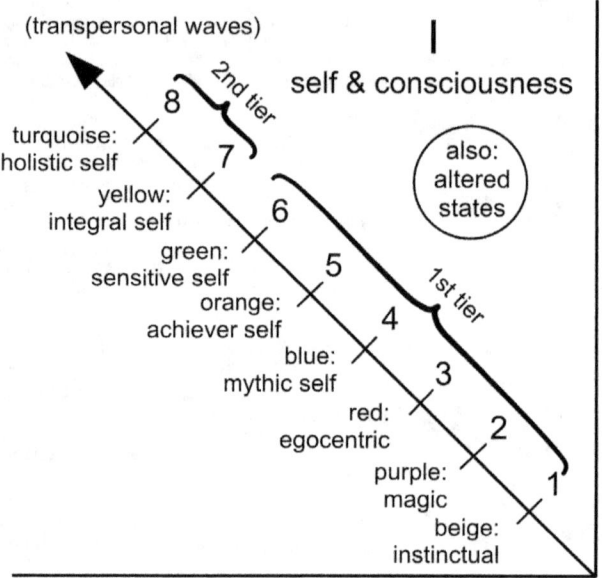

Figure 16.3 Spiral Dynamics and the Individual Subjective (top-left quadrant.) From *The Integral Vision*, by Ken Wilber, ©2007 by Ken Wilber. Reprinted by arrangement with Shambhala Publications Inc., Boulder, CO www.shambhala.com.

Beck introduced a color-coding system to better facilitate communication, allowing the complex collections of attributes associated with each phase to be condensed, labelled, and discussed as a single color. The different labels used for some stages in figures 16.2 and 16.3 reveal alternative perspectives on the same conceptual space. Far from being merely theoretical in nature, SDT found its first major practical application in averting the civil war brewing in South Africa in the early 1990's.[7] During this period Beck travelled to South Africa from Texas sixty four times, meeting with power brokers from all sides. He bore the message that to avert conflict within any society, or between societies, there needs to be an

---

7. Beck at. al, *Spiral Dynamics In Action*, 116-38.

acceptance by society of all developmental levels. This view corresponds with the Integrator-Systemic or Yellow level, which embraces the roles that all preceding levels play. As with the atom's reliance on the quark and proton for its existence, Yellow cannot exist without an integration of all preceding colors. Beck's work in South Africa contributed to bringing an end to the segregationist and oppressive systems of apartheid in 1994.

The role of this integrative process is as important to the individual as it is to the nation-state. The conflict that exists between the unintegrated competing value systems within the individual often predisposes them to addiction. Conflict usually exists between the thinking inherent in any of the first six levels and this conflict can manifest both interpersonally and intrapersonally. On the societal level Beck addressed this in his doctoral dissertation which explored the conditions leading up to the American Civil War. During that period American society was addicted to the way of life afforded by a vassal class. The North had abandoned this narcissistic view, which put the needs of the self above the needs of others, leading to conflict with the south. The result is a subject for the history books, but the addict experiences a similar type of internal warfare daily.

It might be hard to see how a single person can play the role of both parties in such a conflict but, according to Wilber, traumatic events at any step in our development can carve off parts of our consciousness through a process of *splitting*. These parts continue to exist at the level of development where the trauma occurred and become little "subjects"—separate streams of mental activity—that share our own instinct for self-preservation.[8] We could view these subjects as vortices that come into being as the flow of life is disturbed by the damage inflicted on our mental exteriors. They are a symptom of mental turbulence. These subjects do not want to die, but it is hard for us to move ahead with our lives while they maintain their separateness. It is with these subjects that we experience inner conflict. In integral language, this process of splitting is how our shadow parts come into being, making their re-incorporation into our psyche (as illustrated here) one approach to shadow integration.[9]

The existence of these subjects is an energy drain. If we view the amount of developmental energy we are allotted at birth as a collection of one hundred units, it is possible that something traumatic may occur

---

8. Wilber, *A Brief History*, 146.
9. Wilber, *Integral Spirituality*, 119.

at the Red level which prevents its full integration. This will carve off a subject that potentially consumes ten units of developmental energy on a continual basis (through the drag it exerts) until it is integrated back into the self. This can be accomplished by accepting the vital contribution each level plays in making us whole (shadow work.) This integration ends the unconscious conflict that exists between competing internal value systems and frees up the energy that is consumed by that conflict. Using a different analogy, the subject, which continues to drain energy is, as it were, connected to the self by a bungee cord. The forward motion of the self is restrained by this elastic effect so that progression to each successive level in the spiral is impeded.

Add to this the fact that movement from one level to the succeeding level is analogous to a state change in physical matter and another problem becomes evident. For example, large amounts of heat energy are required to change water from a liquid to a gas (steam.) Likewise, a subset of our units of developmental energy are consumed when progressing between successive levels of the spiral. If ten of those units are tied up by a subject at the Red level, there are ten fewer units of energy remaining to power the rest of the climb. This energy can only be released if the Red-level subject is integrated back into the self, at which time it becomes available, once again, to power our upward motion. This is equivalent to fixing the dents and tears in our mental exteriors, caused by trauma, so that life flows smoothly and drag is reduced to a minimum.

Eckhart Tolle and the Holistic Self

An example of this type of integration can be found in the works of writer and spiritual teacher Eckhart Tolle. In his book *The Power of Now* Tolle describes a powerful, albeit unusual, example of such an integration process. He spent most of his early life suffering from severe depression and at the age of twenty nine reached such a point of despair that this suicidal statement arose from within—"I cannot live with myself any longer." After dwelling on this for some time he experienced a profound epiphany—in his own words:

> I couldn't live with myself any longer. And in this a question arose without an answer: who is the 'I' that cannot live with the self? What is the self? I felt drawn into a void! I didn't know at the time that what really happened was the mind-made self, with its heaviness, its problems, that lives between the unsatisfying past

and the fearful future, collapsed. It dissolved. The next morning I woke up and everything was so peaceful. The peace was there because there was no self. Just a sense of presence or "being-ness," just observing and watching[10]

We are all plagued by these mind-made subjects that compose the self, but to experience an immediate and complete integration of all previous levels over the space of a few hours is nothing short of extraordinary. Tolle's epiphany propelled him straight to the top of the spiral, to the Holistic-Universal, where the self dissolves in the realization that ultimately there is only one consciousness of which we are all partakers, a non-dual consciousness where no self/other duality exists. Esoteric and quixotic though this might seem to those of us who have not attained such levels of enlightenment, Tolle is not alone in experiencing this shift.

In *The Power of Now* he describes the weeks following his epiphany spent sitting on benches in Russel Square in central London in a state of bliss. The goal of his follow-up book *A New Earth*, which was selected by Oprah Winfrey's book club and featured in ten educational webinars produced by Winfrey, was to help others achieve this same level of consciousness and help move the world towards the pinnacle of personal development.

Relating with People at Different Levels

As I write this, I feel a sense of embarrassment in recounting Tolle's experience. I currently operate primarily at the Green level (community ecological) but my feelings of bafflement regarding the holistic-universal are instructive. They are the same feelings experienced by someone at the Red level when considering the Orange perspective, or a blue individual when contemplating the Green. We see this every day in news articles about American politics. The Republican Party, which operates at the Blue level (and increasingly the Red) is unable to grasp the Democrats' Orange/Green perspective. The thinking and policies of the Democrats do not compute to the Blue mind. Different parts of the brain and systems of logic manifest in each case. In Canada the connection with SDT in politics is even more noticeable as the colors Blue, Orange, and Green correspond (coincidentally) with the color branding of the Conservative, New Democratic and Green parties respectively.

10. Tolle, *The Power of Now*, 3–5.

The worldviews embraced by each of these parties tend to mirror those of their respective color bands in the spiral.

Politics is always adversarial in nature due to this clash of perspectives, but it is possible to acknowledge the reality and validity of a higher-order or lower-order worldview regardless of your position in the spiral. In the context of Tolle's experience, I am happy to accept the truth of his epiphany without being able to relate to it in logical or experiential terms. The ability to think in this non-judgmental way is something the world needs. This is the second-tier perspective Beck used in his South African endeavors, where a culture existed that accepted the differences between people at all levels in the spiral.

Shame as a Synthesis of Suffering and Trauma

The shadow work of Debbie Ford shows that our inability to reconcile our own nature with the values we have developed over time causes us much suffering. We tend to hate our own limbic attributes, such as hyper-sexuality, tribalism, anger, and aggression (which correspond with the Beige, Purple, and Red levels,) and deny that we are capable of such behaviors or ways of thinking. At the same time these attributes bubble away beneath our calm exteriors and manifest in us from time to time. We attempt to hide them from ourselves and others, but when they are exposed, they promote feelings of guilt and shame. We hate those sensations and bury them using two main strategies: denial and self-medication. (I include in the term self-medication any behavior that results in the upregulation of dopamine activity including food, illicit sexual activity, gambling, and compulsive shopping.) Guilt and shame are the primary "emotional charges" referred to by Ford in her word processing exercise. For example, we read the word "idiot," apply it to ourselves, and view ourselves as "not enough."

The Non-Violent Communication Feelings Inventory lists dozens of feelings, like dread, terror, and horror, that we experience when our needs are not met. These are primary feelings that manifest in the present moment, but when viewing past events, we tend to experience secondary emotions like anger and shame. These are the chronic manifestations of the acute feelings experienced in the past. Regardless of the stimulus in the moment and the feelings experienced, shame is the residual emotion that usually carries over from past experience into the present.

The Beige Hue of Narcissism

Among the more destructive demons are those that come into existence at the sensory-motor or Beige level. At this plane of the spiral there is no concept of language, logic or narrative capacity, and, as elaborated by Wilber:

> . . .the self is largely identified with the sensorimotor world, so much so that it can't even distinguish between inside and outside. The physical self and the physical world are fused—that is, they are not yet differentiated. The infant can't tell the difference between inside and outside—chair and thumb are the same.[11]

If a traumatic event were to occur at this level of development and split off a separate subject as a result, this entity will exert Beige influences in the thoughts and actions of the individual, even in adulthood. As beige fails to differentiate between the self, others, and the material world, it assumes that there is only one perspective that must be shared by all. From this perspective other people are merely objects to be manipulated. To such a person the diversity observed in others is anathema—there is only one view, and that is *their* view. For them, controlling others is as natural as controlling their own arms and other appendages. If they tell their arm to move it moves, so they expect others to act with equal obedience. Their personal preferences also assume the weight of divine fiat. If their reality is the only reality (part of them is still fused with the physical world) then any departure from that reality is an affront to them personally and will stimulate a self-protective reaction. Their perspective is "my view is reality and everyone else is part of this reality" (another example of imago projection.) Departing from adherence to their list of preferences amounts to a rejection of divine law, as there is only one divine reality—the one they have manufactured for themselves. Imagine how confusing it must be for the narcissist when trying to make sense of the non-conformity of others in a cosmos where *they* define "objective" reality.

Getting It Out of My System

I had identified in myself the Beige, Purple, and Red levels as the likely areas where splitting had occurred and determined to integrate these into

---

11. Wilber, *A Brief History*, 144.

my conscious awareness. This decision was precipitated by what could be described as a mid-life crisis similar to that experienced by Eckhart Tolle. I entered this crisis with some knowledge of SDT, thanks to John Dupuy's book *Integral Recovery* [12] and resolved to integrate the Beige, Purple, and Red aspects of my psyche into my Green consciousness by accepting all people, regardless of their level of development, with love.

I achieved my goal through accepting those attributes in myself by loving them in the people around me. In doing so I began my journey from the Community-Ecological (Green) level into the Integrator-Systemic (Yellow) level. This was done in the knowledge that those influences I had been raised to hate were unconsciously a part of me. As shown in figure 16.3, Yellow is the first level in the second tier—a tier at which the strengths and benefits of all prior levels are appreciated. As with our rope metaphor, the rope would not exist without the thread. In the same way, we cannot progress to the Yellow level (a holon consisting of all prior levels) without building on our basic survival instincts (law of the jungle thinking.)

I accepted Purple as manifested in heavy metal music culture, where every beautiful woman is an angel and every talented musician a god, and had philosophical conversations with show girls and First Nations people. I embraced Red in the combat veterans and bikers I met merely from being out in the community. Accepting these people, without judgement, released me from the guilt and shame that came from the disdain I had towards my Purple and Red attributes. In doing this I was following Jung's admonition by "making the unconscious conscious" and was, in colloquial terms "getting it out of my system." I am now free from the tensions produced by my warring attributes and no longer need the negative coping strategies I had used to escape them.

Obviously, I can only provide a cursory treatment of SDT here. You can find a clear and accessible summary of the various levels and their attributes at the *Spiral Dynamic Integral the Netherlands* web site[13] SDT is the work of the late Don Beck, holder of a master's degree in theology and communication, so the underpinnings of SDT include and transcend the traditional Judeo-Christian ethic. Regardless of the scientific/systems template that we overlay on these levels of personal development they are still manifestations of God's handiwork. God created us to accomplish one

---

12. Dupuy, *Integral Recovery*, 41–64.
13. Holwerda, "Home, Spiral Dynamics Integral."

overarching task in life—to move progressively upward through the spiral, both individually and culturally. In the next chapter we will see a Biblical example of personal development on the scale of nation state.

Thematic Takeaways

Feelings: Feelings like dread, terror, and horror that we experience in the present moment may, over time, evolve into secondary emotions like anger and shame. Our addictions are strategies we choose to escape these feelings.

Neural Activities: Our view of the world changes as the neurons of our CNS grow, move into position, create connections with each other and internalize the things we experience into what Steven Peter's labels "the computer" (our memory.) This is the process of growing up or ascending the spiral.

Relationship: Seeing others as objects to be manipulated is a sign of weakness not strength. This behavior is our response to psychic injuries accrued in infanthood, where the only means of establishing comfort and equilibrium was through controlling our members. As the infant fails to distinguish between the inside and outside, those around them are (to this primitive part of the narcissist's consciousness) a part of them to be controlled.

If we approach the relationships we have with others with open hearts and minds it helps us to accept the aspects of ourselves that we have disowned. When we see these disowned attributes in others and accept them without judgement, we are freed to see them within ourselves. Accepting all of our parts is an essential aspect of the recovery process.

What you can do now: Acknowledge that you see others as either stupid or crazy because they embrace a different value system (vMeme) from your own. If you hold Orange values, those with Blue, Red, and Purple values might appear mentally deficient, while those with Green, Yellow, and Turquoise values might seem elitist or delusional. Accept that they are not acting with malicious intent, they are just addressing problems that concern them, but do not concern you. Disagreement need not lead to enmity, and it is possible to appreciate the strengths of each worldview without assenting to their strategies for meeting needs. This is of greatest importance when cultivating self-acceptance. We do not develop Orange values without first holding those of Purple, Red, and Blue. For the Orange subject, accepting these three as the foundation on which the Orange sensibility is built helps calm the turbulence within.

17

Self Actualization

> "The psychology of the mature human being is an unfolding, emergent, oscillating spiraling process marked by progressive subordination of older, lower-order behavior systems to newer, higher-order systems as a man's existential problems change." —CLARE GRAVES

BIBLE HISTORY PAINTS A rich picture of Israel's Spiral ascent in national/cultural terms. The cultural evolution of the nation parallels, in many ways, the stages through which an individual progresses in their transition from child to adult. The Spiral is an upward sweep from egoic/narcissistic thinking to a mindset that integrates and accepts all peoples and perspectives. It initially moves from the Beige view of the newborn to the Purple mystical phase inhabited by the Canaanite peoples of the Levant at the time of Abraham, a level that Jacob and his children also attained prior to their Egyptian sojourn. After a period of 430 years which they began as privileged family members of Pharoah's viceroy and ended in oppression and slavery, God recruited Moses to emancipate Israel from their Egyptian overlords. Hosea describes this in Hosea 11:1 (ESV):

> When Israel was a child, I loved him, and out of Egypt I called my son.

While this can also be read as a prophecy of Christ's return to the Holy Land from exile, its primary application was to the nation of Israel. Israel was God's child, and God cared for them as such, wanting them to mature into a functional Theocracy and leave behind the polytheistic beliefs of the Purple level in which they had been steeped. Sadly, this was a failed proposition, culminating in the loss of their inheritance. The years Israel spent in exile in Babylon commenced with an intervention by God in response to their repeated cycles of reform and relapse. After several centuries in the land, it became evident that they were incapable of overcoming their psychological dependency on idols.

Their experience in Egypt was traumatic to say the least, subject as they were to attempted genocide and servitude. This trauma was a moral injury they soothed through the numbing influence of foreign deities, a strategy to which they had become fused. The exile was akin to a period of rehabilitation from which Israel emerged healed and committed to a stable relationship with their Father. This Blue approach to religious observance set the stage for the coming Messiah and his move away from religious authoritarianism and towards the Orange principles of autonomy and personal responsibility.

Our growth follows the same stages of personal development and comes through the work of institutions, prayer, the accumulation of wisdom, the work of mentors, and the tried-and-true strategies God has made available to us through generations of inquiry. As children of God, we follow the same stages of maturation as Israel, making their story a revelation of God's will for us. This final chapter will explore these issues in some detail to illustrate the way these principles operate in all of us.

Spiral Dynamics and Worldviews

Each level of the Spiral embodies a corresponding worldview and at each level the individual sees their current view as the only valid one. Previous stages are seen as passé and primitive, and those higher up the Spiral as unrealistic, impractical or even quixotic. Equilibria in the demographics of societies usually result in 25 percent of the population sharing the worldview of a base level, 50 percent at the superseding level and the remaining 25 percent at a third.[1] Cultures and individuals tend to manifest the attributes of three levels in roughly those proportions. In twenty-first-century Western society, the twenty-five/fifty/twenty-five split maps to Blue, Orange and Green respectively. Upon entering Babylon, Israel's profile was Purple (idolatry,) Red (warlords,) and Blue (Law.)

---

1. Wilber, *A Brief History*, 134.

Each worldview includes its own set of values. Clare Graves and Don Beck referred to these as value memes (vMemes.) In the Spiral Dynamics context, the term "meme" is derived from an idea introduced by biologist Richard Dawkins in his 1976 book *The Selfish Gene*.[2] Dawkins used "meme" to describe an idea, behavior, or style that spreads from person to person within a culture. In this sense, memes are units of cultural information, akin to genes in biological evolution, that propagate through social learning and communication. As individuals we internalize these memes using lower-left quadrant consciousness. Like genes, the cultural memes that result from trauma are passed between generations (generational trauma,) a principle that may lie at the root of Israel's fusion with Purple values and strategies. The following sections examine six vMemes in their Biblical setting and how they relate to maturation in both individuals and cultures.[3]

The Purple vMeme

The worldview of both the (Purple) Canaanite and Jewish people in Egypt was:

> The world is a mysterious and frightening place; it is controlled by mystical spirit beings that need to be satisfied and pleased.

Israel's struggle in their egress from Egypt and throughout the period of the Judges, was to draw on the strength of Red and resist the magnetic pull of Purple. During this time, they alternated between faith in the Lord of Hosts (armies) and idol worship. The difficulty they had in committing to God and God's system of laws and regulations (Blue) illustrates the huge amount of energy required to move from one level to the next in the Spiral. This population, 25 percent of which ascribed to the Purple worldview, 50 percent to Red and 25 percent to Blue, progressed through many cycles of sin, supplication, and salvation, falling back from Red to Purple each time, before being propelled by the suffering of God's judgements to see the value his Blue Law embodied and asking him for help. While belief in God requires faith in a mystical spirit being, God was not interested in stagnation and wanted Israel to

---

2. Dawkins, *The Selfish Gene*.
3. Holwerda, "Home, Spiral Dynamics Integral."

climb the Spiral, individually and communally, embracing faith in the one true God, as defined by Law.

The Red vMeme

On leaving Egypt, Israel's driving principle became Red, where:

> The world is a jungle. Survival of the fittest. "I am taking charge without taking others into account."

This remained their center of gravity throughout the time of the Judges, a mindset that empowered them to wrest the Holy Land from the Canaanite peoples. If Israel had taken the feelings and needs of the Canaanites into account, God's plan for them would not have been realized. A Red society consists of clans or tribes led by warlords and dictators. We see this in the twelve tribes of Israel and leaders like Joshua, Caleb and Ehud.

The values associated with each vMeme flow from current life conditions. Increasing stability in life conditions at one level forms the foundation on which to build the next. If that foundation subsequently loses its integrity, because life conditions experience a regressive change, the culture will move back down the Spiral to a position where values and beliefs best match the needs of the moment. If Red's attempts to take charge prove ineffective, falling back to Purple provides the tools needed to enlist the help of mystical spirits in the hope that they will prevail where personal power failed.

The Blue vMeme

When King Solomon acceded to the throne, Israel's center of gravity moved into Blue and a lengthy period of peace ensued. It was Israel's request for a king that drove the metamorphosis from Red to Blue. Their desire was for an authority figure who would impose order on their existence, something that was missing under the chaos of Red. The essence of Blue is:

> Order, authoritarianism, Holy, meaningful.

Each level in the Spiral confers advantages over the previous level but also introduces challenges. Individuals are propelled by a dissatisfaction with the status quo towards a new paradigm that provides solutions to the limitations of the present. However, the solutions of the present often become the problems of the future. Following their request for a king, God delivered a warning in 1 Samuel 8:10–18. This passage shows that while monarchic rule provides societal unity and stability, it has its own set of limitations centered around personal autonomy. A king, God says, will take from you what you value and use it for his own gain. The individual's lack of control over their quality of life is the primary deficit that Orange seeks to address. This is seen in Marxism's coopting the means of production from the bourgeois overlord and the corporation's resistance to government regulation. In the Old Testament it was briefly manifested in Israel's rejection of Rehoboam's authoritarian overreach when he asserted in 1 Kings 12:10 (NIV) "my little finger is thicker than my father's waist."

The period of the kings followed a similar trajectory to that of the judges in that aspects of Purple continued to exert their influence. While Solomon's reign subsumed Red by establishing a decades-long time of peace, this came at a cost. To achieve détente, Solomon made treaties with the surrounding nations, often intermarrying with the royal families of those nations. As 1 Kings 11:4 (NIV) reveals:

> As Solomon grew old, his wives turned his heart after other gods, and his heart was not fully devoted to the LORD his God, as the heart of David his father had been.

This Purple influence on the nations of both Israel and Judah was ultimately vanquished by the Babylonian exile, during which they served the rulers of other nations rather than their own. After returning to their land and constructing the second temple, they had reached a state of maturity where the allure of idols no longer held them in its thrall. The political environment made it impossible for them to appoint their own king, but the baton of authoritarian power passed on to their religious leaders. During this time Blue (with its traditional worldview) was the governing ethic.

The Orange vMeme

At the culmination of this authoritarian period, Israeli culture was ready for the next level—Orange. Orange embodies the modern worldview, a mindset that has sprouted on occasion throughout history, but which did not fully bloom until the current epoch. The key feature of this color band is individual autonomy. This showed itself in Christ's adversarial stance towards authoritarianism, shown in his resistance to the religious leaders of the time. The success of Christ's ministry was, in large part, due to the willingness of the culture to set aside Blue and the Red influences. While SDT identifies the accumulation of wealth as the preoccupation of Orange, this in no way clashes with the teaching of Christ. Christ implored his hearers to put aside the worldly wealth of materialism and embrace the true riches of God. As Paul writes in Ephesians 3:8 (NASB):

> To me, the very least of all saints, this grace was given, to preach to the Gentiles the unfathomable riches of Christ.

The Orange worldview is summed up as:

> The (makeable) world is full of chances and opportunities, and the world can be fully understood by using rational thinking.

While the book you are reading draws from all six first-tier levels to guide us on our recovery journey, its primary focus is the integration of Orange and Green—the Orange belief that the world can be fully understood using rational thinking, and the Green view that our social and environmental structures are fragile and we need to give and receive empathy (with ourselves and others) to attain peace and the unity required to build contentment. In Matthew 23:8 (NIV) Christ establishes the bounds of authoritarian power in his ecclesia:

> But you are not to be called 'Rabbi,' for you have one Teacher, and you are all brothers.

Trade unions embrace this spirit when one member acknowledges another as brother or sister, a practice that works well in an Orange culture but is viewed by Green as the promotion of binary gender stereotypes. This language contrasts the authoritarian worldview with one of equality. With Christianity there was still an ultimate authority, the one who Christ refers to in this verse as "Teacher" (himself,) but his authority exists beyond the realm of the Spiral. The Spiral is encoded in our social DNA, while Christ's authority transcends the material order.

From this point forward, the tribal and monarchic principles that had formed the basis of Israel's social framework increasingly lost their relevance, opening the door for the Gentiles' adoption into *spiritual Israel*. If the religious order had continued using strict us-and-them terms, the word of God would never have spread beyond the Jewish community to encompass all of humanity. It was the progression through Purple,

Red, and Blue that made this possible. The proclamation of God's word to the Gentiles was also empowered by the entrepreneurial spirit of Orange—the ability to grasp hold of opportunity.

It was this entrepreneurial spirit that powered Paul's mission to the Gentiles. Not only did he support himself financially using his own tent-making business, but in spiritual terms he was a billionaire. The entire globe was enriched beyond measure by the abundance of Christ's grace at work in him. What greater investment could Christ have made in one individual than that bequeathed on the road to Damascus?

Integrating Lower Levels

Following the establishment of the church, external influences re-established the authoritarian power structure of Blue. The Christian church also returned to the mystical roots of Purple by absorbing aspects of the pagan festivals of Saturnalia (Christmas) and Easter, which celebrates the Goddess of the radiant dawn, of up-springing light (Ishtar) of Babylonian provenance. The extent to which these belief systems were conflated with the teaching of Christ is debatable, but maintaining traditions that correspond temporally with those already extant would broaden Christianity's appeal.[4] Red (the appropriation of resources) and Blue (the imposition of order) were exhibited in the violent excesses of the Crusades and the Spanish Inquisition. Blue also found substance in the supremacy of the Roman Pontiff and the hierarchy of cardinals and priests whose ceremonial garb and religious practices mirror those described in the Pentateuch. All of this was a necessary "evil," an unavoidable pre-requisite for the acceptance of Christianity by the adherents of the four worldviews expressed in early Christian culture. While these developments were not a true societal manifestation of Yellow thinking (something that remains only embryonic in our twenty first century cultures) Yellow logic underlies much of what transpired in the early Christian era. The early Christian leaders brought about an integration of Beige, Purple, Red, and Blue within the church.

While each of these perspectives sullies the purity envisioned for the church by its founders, their inclusion was an unavoidable manifestation of our human cultural DNA at work. This was an embryonic expression of the Yellow worldview where:

---

4. Woodrow, *The Babylon Connection*.

> The world is a complex, self-organizing, natural system that requires integral solutions.

We all require integral solutions to heal the hurts of our past. The addict seeks release from the warring influences of each trauma-dissociated level as it manifests within. Their evolving selves pursue emancipation from the shame rooted in past deeds, when in reality those things were done *to* them, not *by* them. What we all seek is serenity, but all too often by chemical means. Our work is in finding the serenity that flows from accepting our own past and developing the latent strengths that lie within. In the words of the Serenity Prayer:

> God grant me the serenity to accept the things I cannot change; courage to change the things I can; and wisdom to know the difference. Reinhold Niebuhr (1892-1971)

The Green vMeme

But what of the Green? The Green worldview is:

> The world is a community for the whole of humanity who shares Mother Earth as their home.

While aspects of each worldview are manifested in individuals of all eras, a new worldview does not emerge at the societal level until the limitations of the dominant view become evident to many. According to SDT wisdom, the Orange vMeme emerged as a society-wide influence 300 years ago, Green 150 years ago, and Yellow fifty years ago.[5] This does not mean that Orange, Green, and Yellow thinking were absent prior to these epochs, merely that their corresponding worldviews were not activated in large segments of society.[6] Without Orange (modern) thinking[7], the gospel message would not have been effectively promulgated, without Green cognition (post-modern) Christ's message of love, forgiveness, selflessness and inclusivity would have fallen on deaf ears and without the integrative principles that are fully expressed at Yellow, the big-tent plan of the universal, inclusive (Catholic) church (with all its faults) would have failed.

---

5. Beck et al., *Spiral Dynamics in Action*, 75–76.
6. Wilber, *A Brief History*, 127.
7. Küstenmacher et al., *God 9.0*, ch. 5.0.

It is interesting to note that while Catholicism was able to embrace the first four vMemes, Orange is almost completely absent from the value system of the Holy See. This is the great challenge of all religious organizations—reconciling the magical beliefs of Purple (where the God of Spirit is most relatable) and the materialistic/rational perspective of Orange.[8] Some denominations exercise rationality through the study of history, archeology, and Bible prophecy seeing these as lines of evidence that validate the Biblical account, while others find a different path through Orange. Orange extremists completely deny the validity of the Bible and everything it contains. While they might accept the existence of Jesus as a historical figure, they see historical and archeological accounts as the only sources of truth in the matter. For this reason, Orange tends to form a glass ceiling which prevents many religious organizations from progressing beyond holy, authoritarian Blue.

The emergence of Green thinking at the societal level is a response to the failure of Orange to account for the complexity and fragility of the natural order. Orange-focused thinking seeks to maximize the effectiveness and productivity of the individual, allowing them to create their own world of comfort and accomplishment. This is exemplified in the drive to elevate agricultural productivity following the Second World War. Orange thinking sees the progression of science and technology (the rational world) as the be-all and end-all of human accomplishment, and the abandonment of Orange technologies as a regressive act. Yet the effects of chemical fertilizers and pesticides on the environment are an issue that needs to be addressed; these technologies are Orange solutions to yesterday's problems that have become progress traps in the present.[9] While the preoccupation of Orange thinking is to better the living conditions of the individual, environmental destruction deteriorates the conditions required to create ease, comfort, and ultimately, wealth.

This situation pits short-term against long-term thinking and the interests of the individual against the long-term good of the community and Mother Earth. While Orange thinking gives lip-service to the equality of the individual, this equality is couched in narcissistic terms—how can I maximize my own advantage within a community of equals. Green thinking is based on pragmatism, caring, and self-abnegation, and promotes the good of the community over the good of the individual. It finds

---

8. Wilber, *Integral Spirituality*, 179–82.
9. O'Leary, *Escaping the Progress Trap*.

expression in social programs, universal health care, guaranteed basic income and the immigrant-based economy. In this, the principles of nationalism, tribalism, and racial judgement are discarded. The current trend towards populism illustrates how Red, Blue, and Orange always struggle to survive in the face of progressive thinking. Progress, in this sense, being the continued climbing of the Spiral.

A defining feature of Green thinking is the ability to see things from the perspective of the other, be it the poor, the refugee, or those persecuted by the Blue principle of uniformity. Blue thinking longs for order and demands conformity to societal and sexual norms. Green allows for the expression of all forms of being because they are inescapable components of reality, the denial of which hurts those on the periphery. The Green mind hates the sight of others' suffering because it activates a pain response within the observer. For this reason, the acceptance of homosexuality and trans-genderism is an inescapable component of the Green vMeme. This pits the Green ethics of empathy, diversity, and inclusivity against the Blue axiom of cis-gender conformity.

I raise these issues in the context of addiction to show that the conflict between the different vMemes within an individual generate similarly "charged" emotional dynamics and their associated dysphoric states (shame, self-loathing, and anger.) SDT provides another perspective from which to view our shadow elements (as discussed in chapter 15,) exchanging the focus on words as emotional triggers with a consideration of conflicting value systems. For the individual these conflicts can manifest on an unconscious level by their denying the more primitive vMemes within their own psyche. Benefit can be gained by accepting the *strengths* exhibited by the healthy expression of each value system. Without re-integrating the trauma-dissociated subjects of our past and calming our mental turbulence, the dysphoria persists along with the destructive strategies a person has devised as a means of escaping it. To be psychologically whole and build resilience against our addictive tendencies we must accept the contribution each level has made to our lives.

An Integral View—Modern-Day Israel through a Yellow Lens

Setting aside the spiritual Israel of New Testament provenance for now, there is much to learn about SDT and the contributions of each level in the Spiral from a consideration of the modern Jewish State. Following the Bar Kokhba revolt (AD 132-135,) Israel was again subjected to exile and

SELF ACTUALIZATION

their people were dispersed throughout the world.[10] In 1948, following the horrors of the Holocaust and under the mandate of the Balfour Declaration of 1917[11], the United Nations supported the establishment of a National Homeland for the Jews in Palestine. This laid the foundation for the return of the Jews to their former homeland, but not without introducing the troubling political dynamics we read about in the news daily. A second-tier consideration of the Israeli Palestinian relationship illustrates the benefits that accrue from accepting all first-tier value systems as they are revealed in ourselves and others. It also shows that without the progression of leadership to the second tier, such levels of acceptance are impossible to achieve. The following discussion of the Israeli Palestinian dynamic is meant as an archetypal consideration of the tensions between value systems that might exist within the addict.

SDT in Israel 2004–2008

In the mid 2000s SDT proponent Don Beck and his team began applying SDT principles to the Israeli Palestinian conflict.[12] One of the preliminary tasks of this project was to establish a value-system profile for each community. These were used to assess the best strategies for communicating and reasoning with the powerbrokers of each party. Figure 17.1 below utilizes a two-dimensional representation of vMeme distribution within the two cultures.

---

10. Menahem Mor, "The Second Jewish Revolt."
11. Al Tahhan, "More than a Century."
12. Beck et al., *Spiral Dynamics in Action*, 139–88.

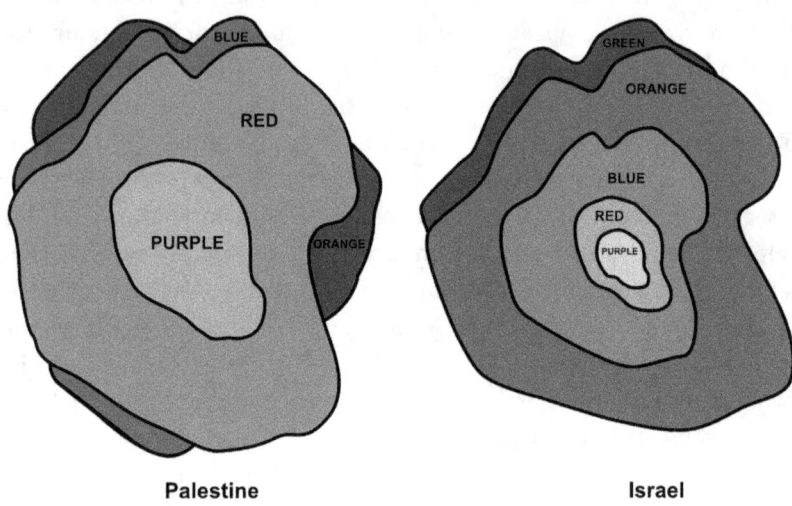

Figure 17.1 The vMeme Profiles of Palestine and Israel. Used with permission from admin. of Elza Maalouf's estate Said E. Dawlabani and the Center for Human Emergence Middle East. https://www.humanemergencemiddleeast.org.

Very few Israelis ascribe to the Green vMeme, a color that is completely absent from the Palestinian value-system profile (meme stack.) The prevalence of Red (the drive to fight for our own interests) in the Palestinian profile is notable, as is the dominance of Orange (science, technology, and entrepreneurship) in the Israeli profile. The Palestinian profile shows that a culture cannot develop robust Blue or Orange cohorts when they are fighting to establish a nation state. In the 1930's and 40's the Jews were striving with Great Britain for control over parts of Palestine and at that point in time their vMeme profile resembled that of the Palestinian state today. This illustrates the influence life conditions have in shaping dominant cultural values. Additionally, a culture cannot have Blue government institutions or successful technology or manufacturing sectors without land to build them on.

Although Beck saw much enthusiasm for developing Orange industries in Palestine, the state lacked the robust Blue institutions required to guide their development. This was a major impediment to developing common ground between the two parties, as Israel's interest in working with Palestine was predominantly framed using Orange themes. The inability of Palestine to develop a robust Orange cohort illustrates the reliance Orange has on a Blue foundation. Endowing any Orange group with a second-tier perspective of Blue would help them acknowledge their dependence on Blue institutions for their own success. While Orange spurns authoritarianism, without a Blue authoritarian foundation, Orange consciousness is a non-starter. Tensions in the addict's relationship with the Blue values expressed by authority figures in their upbringing can impede progress to Orange and might cause a regression to Red.

Tailoring the Approach to a Cohort Based on Their vMeme Profile

Beck showed that providing aid to the Palestinians using Green language and values was a failing proposition. This is because the life conditions in Palestine are almost entirely Purple, Red and Blue and demand a response that people at those levels find relatable. As described by Elza Maalouf[13] the services provided by Non-Governmental Organizations in Palestine were predominantly Orange and Green, which clashed with Palestine's Purple and Red value systems. This clash of values rendered most of these programs ineffective and unsustainable in the Palestinian context.

This emphasizes John Dupuy's assertion that addiction counseling be provided using the worldview language of the individual seeking support. Any deviation from this approach results in counterproductive levels of cognitive dissonance. Dupuy uses Spiral Dynamics and Wilber's AQAL matrix to tailor his treatment programs to his clients' unique needs and life conditions. To be clear, Purple and Red thinking is neither the basis for nor a symptom of addiction, but a response to life conditions. There is nothing inferior or objectively wrong about them, their emergence is baked into our cultural DNA as the most appropriate response to a particular environment.

Green Versus Red

Red consciousness not only embraces the use of violence and force but views these as the most efficient and effective solutions in most situations. To the Red mind, there is nothing morally wrong in their expression and those who utilize them are viewed as having determination and strength of character. They are the mighty heroes of the group and might attain the standing of warlords or tribal leaders. They are freedom fighters rather

---

13. Beck et al., *Spiral Dynamics in Action*, 161.

than terrorists. In contrast, the Green vMeme sees the use of violence as totally counter-productive and morally wrong. Green considers more situational complexity and ignores short-term solutions in preference to long-term ones. Programs that propose pacifist solutions to a culture that sees violence as the best solution are bound to fail.

Short-Term Versus Long-Term Solutions

An example of the short-term fix is the destruction of the homes of Palestinian "terrorists" by the Israeli Defense Forces (IDF) in response to Palestinian attacks on the state of Israel. This is a Blue response to a perceived Red infraction. The IDF sees this as an effective deterrent against future offenses of the same kind, while Green views such tactics as a cause of increased alienation between the two sides that draws more people into the fray and provides even more impetus for exacting revenge. The Blue solution, in the Green view, is bound to devolve into a never-ending spiral of tit-for-tat. A Green response would be to foster empathy between the two sides and promote an understanding on the part of one for the grievances of the other. After each side accepts their opponents as people with feelings and needs, rather than demonizing them, tensions should diminish.

Red consciousness will likely see the Green proposal as morally wrong, displaying a lack of loyalty to the band, a sell-out and tantamount to treason. This tension between Red and Green led to the 2005 assassination of Israeli Prime Minister Yitzhak Rabin by an ultra-nationalist, immediately following a peace rally held in support of the (Green) Oslo Accords. His assailant, Yigdal Amir, embodied the settler-Red ethic. He agreed with many Israeli conservatives who saw the accords as an attempt to forfeit the occupied territories and capitulate to Israel's enemies.

On the individual level, if, because of traumatic events experienced at the Red phase of development, a Red *subject* has split from an individual's identity and continues to exist in their subconscious, the same kinds of tensions can wreak havoc as their conscious worldview develops through subsequent levels in the Spiral. Such individuals might seek to escape these tensions using addictive substances and behaviors. The potential for such a condition should be considered as part of every addict's recovery journey, the goal being to integrate an acceptance of healthy Red consciousness back into the self.

Moral Relativism

Notice that in the context of the conflict between Red and Green, morality is relative. Green morality and Red morality are in fact, in this case, mutually exclusive. This indicates that much of what people of faith label as "moral relativism" is a clash of worldviews. Each level in the Spiral has its own sense of right and wrong. The word of God itself hints at God's consciousness of this principal. He called his Purple son out of Egypt, using Red language and tactics to establish a Jewish homeland, while laying the foundation for the Blue kingdom that followed. In His role as Yahweh Tzevaot (the Lord of Armies) He acknowledged that might is right. Christ, in contrast, rejected the use of force (all who draw the sword will die by the sword—Matthew 26:52) instead preaching the Orange values of self-determination through faith and the Green ideals of self-abnegation, compassion, and love. To Christ, might was definitely not right. These stances are not contradictory, but rather the most appropriate responses to the life conditions of the time. When Christ's Green message fell on Red ears, or the Blue ears of the contemporary Jewish leaders, those ears failed to hear.

The growth of God's son (Israel) into an acceptance of Orange and Green values was a pre-requisite for the continuation of God's purpose through Christ. The tension between Christ's teachings and those who refused to hear them stems from the fact that every "first tier" stage (Beige thru Green) rejects the benefits of each stage lower down or higher up the Spiral. Orange is the enemy of Blue, Blue the enemy of Red and so on. It is not until the second tier vMeme, Yellow, emerges that the true value of each level can be comprehended.

On the individual level, tensions between conflicting value systems activate the amygdala, where the *guard dog* fails in its attempts to serve two or more masters. Chronic activation of the amygdala, in this case, can lead to self-structure disintegration—the collapse of the spiral all the way back to beige. This is the rock-bottom of the addict, where the only law is that of the jungle and the next fix is the only need. Disempowering the negative charge of our baser attributes through an acceptance of their value helps to replace our fragmented state with an integrated one, a structure with a single unified core.

The Personal Benefits of vMeme Integration

Embracing the Red attributes of my psyche has certainly reduced internal conflict and allows me to feel at ease with the ex-convicts I associate with in my recovery work. Just last night I was locked in a warm embrace with a friend who has a criminal record three pages long, including one count of second-degree murder. His behaviors, which are viewed as crimes through the Blue lens of law enforcement, are understandable in the light of Red's value system—"I am taking charge without taking others into account." His behavior is further explained by considering his upbringing. His father was a police officer who tried to *impose* his own imago projection of what a good son should be. One tactic his father used, to this end, was forcing him to clean police cars for free. In my associations with Celebrate Recovery, it is always uplifting to see Christ's willingness to influence the lives of such men for good and the willingness they show to be transformed.

Although the Green vMeme embodies first-tier consciousness and generally avoids integration with other first-tier levels, Non-Violent Communication (a Green technology) does include a Blue component. The first time I heard about NVCs principle of "the protective use of force" I saw it as a complete contradiction. How could the use of force be compatible with Non-Violence? This reaction, which was based on emotion as much as it was on reason, is the influence that promotes the heterogeneous nature of the first tier. A strength of second-tier consciousness is its ability to make sense of paradox. From the NVC perspective the use of force might be perceived as evil, but in certain situations it is necessary. The amalgamation of Blue authoritarian principles with the pacifist Green ethic is an example of shadow integration. This kind of reasoning was embodied in Albert Einstein's Green-inspired push for a single World Government which would maintain a monopoly on the use of military force. In 1932 he produced an article for a meeting of the League of Nations, Russia, and the US in Geneva in which he wrote:

> Mere agreements to limit armaments confer no protection. . .there should be an international body empowered to arbitrate disputes and *enforce* the peace.[14] My italics.

Speaking as a WWII era Jew and ardent Zionist, he had personally experienced the use of military force in the expropriation of his own private

---

14. Isaacson, *Einstein*, ch. 16, "Einstein's Pacifism."

property. The use of force is a strategy of every level in the Spiral and is not limited to Red; the distinction between the use of force by Red, Blue, Orange, and Green is the value system that motivates its expression. Each level will use force to protect the things it values. It is a testament to the complex intersection of international politics and human nature that the downtrodden Zionist has transformed into the authoritarian overlord in their attempts to enforce peace in their homeland.

An Integral View of Modern-Day Israeli Society

A second-tier overview of Israeli society is shown in figure 17.2. This figure shows that each vMeme has its own function based on its intrinsic strengths. In this map, the vMemes are distributed geographically based on living conditions. The "settler Red" classification is the most controversial due to the Green leaning of most Western governments and their disdain of Red goals and methods. They do not want citizens of a close ally to appropriate land in another jurisdiction. Western countries see Blue authoritarianism as a necessary last resort, hence most of them maintain sizeable armed forces while Israeli settlers see the land as theirs by divine fiat and will employ Red methods whenever they please. The primary role of the Israeli Defense Forces is found in the Blue functions of policing, intelligence, border control and deterrence.

The Orange of Israel's high-tech industries and highly technical horticulture sector forms a buffer between the Blue and coastal Green, completing their gamut of first-tier vMemes. No culture, as of this writing, has achieved second-tier consciousness but to the addict, the sense of integration, unity, harmony, and community connection that exists on the yellow plane is of great value in the recovery process.

300   PART THREE—*INTEGRATION:* SUSTAINABLE RECOVERY PRACTICES

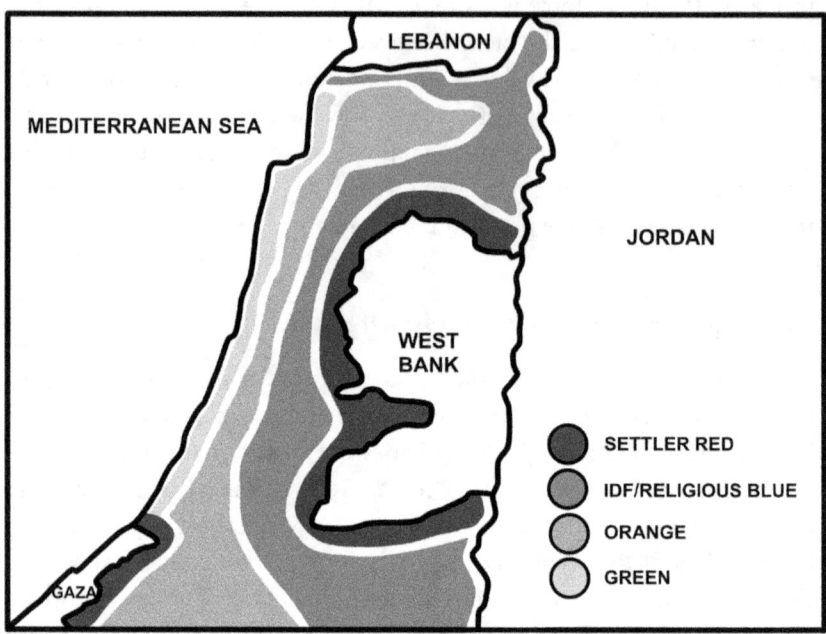

Figure 17.2: A Geographical View of vMeme Distribution in Modern Israel. Used with permission from admin. of Elza Maalouf's estate Said E. Dawlabani and the Center for Human Emergence Middle East. https://www.humanemergencemiddleeast.org.

A Summary of Human Maturation (Growing Up)

vMemes emerge in the individual as their life conditions progress from simple to complex. These conditions progressively unlock latent developmental capacity in the individual. They usually exhibit values commensurate with their current level of development and spurn the values held by those at other levels. A culture consists of individuals at various levels of development and operates at the level attained by the majority of its population. Each individual gravitates towards communities and occupations that match their system of values. In the same way as an atom depends on the quark, proton, neutron, and nucleus for its existence, so Green depends on Orange, Blue, and Red. The lower levels are the foundations on which higher levels are built.

Sadly, Beck's early twenty first century endeavors in the Middle East were brought to an end by the financial crisis of 2008, during which many of his sponsors experienced financial difficulties. Without a steady flow of funds, he had no choice but to withdraw, handing over control of the project to the United Nations. Without his leadership, however, the project bore little fruit. How different the newspaper headlines might be today were he able to bring the venture to completion.

The Role of Self-Actualization in Addiction Recovery

Abraham Maslow and Clare Graves taught contemporaneously at Union College in upstate New York during the 1940's and early 1950's. During Maslow's leave of absence from 1948 to 1949, Graves acted as a substitute, picking up where his friend had left off in his teaching responsibilities and research. This was the period during which the ideas behind his emergent cyclical theory of biopsychosocial development (the precursor to SDT) began taking shape. As Maslow's hierarchy and SDT both model the same psychological space, it is not surprising that there have been attempts to harmonize the two perspectives. Some take a linear approach, which maps Beige to Maslow's first level (Physiological Needs) and Graves's second-tier to the highest level of Maslow's hierarchy (self-actualization.) While there is some congruence in this approach, it denies those at any of the first-tier levels the ability to attain to the highest level in Maslow's hierarchy.

Actualization is defined as *bringing something into being or making it real*, but how does this apply to the self? Think of this in terms of the apprentice carpenter. During their apprenticeship they work under a journeyman or master while their skills are being developed. Eventually they themselves attain the status of journeyman and may progress to master should they so wish and if their career affords them the opportunity. A journeyman carpenter is a *real* carpenter, but they have not yet attained the pinnacle of their capabilities. They are self-actualized, at the level of journeyman, but there is still much to learn and much maturation to undergo.

While the levels of development described by SDT can be harmonized with those of the hierarchy in that they both move from basic survival instincts to a state of psychological maturity, Graves noted that aspects of self-actualization exist at each level of the Spiral.[15] An individual who recently progressed from Purple to Red, for example, does not at that point fully embody Red consciousness. It takes time and experience

---

15. Allen, "Self-Actualization."

for Red to come into being and for them to actualize Red consciousness. It is at the point of actualization that the limitations of Red become apparent, and the individual begins to see Blue solutions as remedies for the deficits of Red. At this point they become an apprentice Blue subject. This process is repeated at every level in the Spiral.

While attaining second-tier consciousness could be described as a state of actualization, the expression of Yellow (the first level of the second tier) in the individual is still merely a stepping stone on the Spiral. After the individual becomes actualized at the Yellow level, Turquoise solutions to Yellow limitations begin to present themselves. This cycle can continue indefinitely. While we may be enlightened in our current level of thinking, we are still naïve to the attributes of higher levels that are adept at handling more complex living conditions.

At each level, we progress through the final four stages of Maslow's hierarchy (physiological needs are inherited from the previous level,) progressing from safety to belongingness and love, then to esteem before attaining a state of actualization. If our life conditions are Red in nature, there is no reason to proceed to the next level. Self-actualization at Red is all we need to be real authentic individuals, having reached a state of maturity that requires no further augmentation. However, if we define maturity as the willingness and ability to adapt to ever more complex living conditions, the maturation process can be seen as essentially limitless. We can find maturity at each level in the Spiral but still have a lot of growing up to do.

At each level, except Beige where there are no preceding levels, the addict may be plagued by dissociated subjects from any of the prior levels. For the Red individual, the integration of Beige and Purple subjects into their Red consciousness provides sufficient structure for the psyche to exist in a state of grounded equilibrium. This is why SDT-based addiction counseling is tailored to the level extant in the individual being mentored. Red consciousness needs help in becoming authentically Red rather than being dogged by dissociated Beige and Purple subjects. The same applies to those at all remaining levels in the Spiral.

The Expression of Dopamine Throughout The Spiral

Resistance to growing up could be described as the Peter Pan syndrome. We resist it because at our current level of development, there is no reward in contemplating the values and activities of subsequent levels and no incentive to engage with them. We view higher levels as a threat to our psychological wellbeing, which activates our amygdala and creates a state of aversion. Not only are higher levels regarded as boring, but they are seen as impediments to expressing the values and behaviors that provide succor and reward in the present. These values and coping strategies are the only intrinsic motivators we can conceive of at our current level.

As a Purple teenager, I remember how outlandish the Orange, middle-aged adults appeared as they discussed the performance of their investment portfolios. If this was what growing up looked like, I told myself, I want none of it! The things that engaged me at the time were hard rock culture and girls, the dangers of which I was regularly warned of by my Blue parents. The association of rock culture with drugs and promiscuity were influences that activated the *guard dogs* of my parents. They in turn obtained their reward by exercising their authoritarian prerogative over their adolescent son.

And that is the great thing about growing up. At each level there *is* a reward, it is just activated through different means. Additionally, those things we found rewarding at lower levels loosen their allure as we progress up the Spiral. Using Wilber's quadripartite strategy of wake-up, grow-up, clean-up, show up, individuals at any level in the Spiral can experience the most fulfilling natural dopamine high by putting aside the counterfeit rewards of their destructive coping strategies and showing up as their real, authentic selves in service to the world.

As you are still with me, thank you for your perseverance. I hope you were able to resonate, on some level, with my experience as a unique member of humanity, despite our myriad differences and the paradoxes we sometimes embody. I hope that reading this book has also been meaningful for you. There is nothing more meaningful to me than to contribute to the well-being of others and help them navigate the stormy waters of this vast ocean of consciousness. I will sign-off, sending you my best wishes for your recovery journey (and we are all recovering from something) in the words of Numbers 6:24-26 (NIV):

> The Lord bless you and keep you; the Lord make his face shine upon you and be gracious to you; the Lord turn his face toward you and give you peace.

Thematic Takeaways

Needs: The individual's ascent of the spiral is driven by needs. At Beige, needs are oriented around subsistence and security. Purple introduces the need to connect with other individuals; Red embodies the desire for power and Blue quenches the longing for order and structure. The needs that define the next level are not perceived as imperatives until the limitations of the current level have become apparent.

Neural Activities: There is no association in Spiral Dynamics between individual vMemes and a particular set of brain regions, but I might be so bold as to suggest that the territory expanding drive of Red is primarily a function of the amygdala, the capitalist impetus of Orange relies on the anticipatory role of the nucleus accumbens and Green's compassion, goodwill and caring for the ecosystem stem from the activities of the prefrontal cortex. Establishing observational evidence for these assertions would contribute greatly to the study of the field.

Relationship: The father/son relationship between God and the nation of Israel illustrates the benefits of structure and responsibility in the life of the addict. God was as actively involved in their life challenges and growth towards maturity as he is in ours. We see ourselves as sovereigns (in control) of our lives, which can work to both our advantage and our disadvantage. When we use our power in destructive ways, our mentors play the role of God's prophets in the kingdom of our lives. Mentors are not subject to *our* limbic drives, and this makes them suitable conduits for the wisdom of God. It is through the relationships we have with God and the people he places in our lives that a new, mature version of ourselves comes into being.

What you can do now: Reading this book has been quite an accomplishment! I hope you were able to see your true appearance reflected in its pages. Do not waste your investment of time and energy by forgetting what you look like! (James 1:22-25) Above all else, find people with the skills to guide and support you through the momentary challenges you face. Attend recovery groups, make friends, find a sponsor or recovery coach. Do not just be a reader, but also a doer, walk the talk and put all your faith and trust in that higher power who gives structure to your frame.

Bibliography

aans.org. "Parkinson's Disease – Symptoms, Diagnosis and Treatment." Aans.org, 2019. https://www.aans.org/en/Patients/Neurosurgical-Conditions-and-Treatments/Parkinsons-Disease.

ABC. "Murray Trial: Lawyers Argue Jackson Suffered from Demerol Withdrawal." ABC News. Sep. 29, 2011. https://abcnews.go.com/Health/conrad-murrays-attorneys-demerol-withdrawal-caused-insomnia-led/story?id=14633134.

Al Tahhan, Zena. "More than a Century On: The Balfour Declaration Explained." www.aljazeera.com. Aljazeera. Nov. 2, 2018. https://www.aljazeera.com/features/2018/11/2/more-than-a-century-on-the-balfour-declaration-explained.

Allen, Lindsay. "Self-Actualization and 1st Tier of Spiral Dynamics." Medium, Jan. 4, 2022. https://medium.com/@lallen_66499/self-actualization-and-1st-tier-of-spiral-dynamics-c2d35ed567fd.

American Psychological Association. "Eye Movement Desensitization and Reprocessing (EMDR) Therapy." *American Psychological Association*, 2017. https://www.apa.org/ptsd-guideline/treatments/eye-movement-reprocessing.

Arehart-Treichel, Joan. "Don't Blame Amygdala for Meth Users' Aggression." *Psychiatric News* 45.24 (Dec. 17, 2010): 27–27. https://doi.org/10.1176/pn.45.24.psychnews_45_24_025.

Ariely, Dan. *Payoff: The Hidden Logic That Shapes Our Motivations*. New York: Simon & Schuster/TED, 2016. Kindle Edition.

The Attachment Project. "Attachment Styles & Their Role in Our Adult Relationships." Attachment Project, Jul. 2, 2020. https://www.attachmentproject.com/blog/four-attachment-styles/.

Baer, Drake. "Apple Has Reportedly Stopped Banning This Mindfulness App from the App Store." Thrive Global, Apr. 12, 2017. https://medium.com/thrive-global/apple-has-reportedly-stopped-banning-this-mindfulness-app-from-the-app-store-e712e83d90e5.

Báez-Mendoza, Raymundo, and Wolfram Schultz. "The Role of the Striatum in Social Behavior." *Frontiers in Neuroscience* 7.233 (Dec. 9, 2013). https://doi.org/10.3389/fnins.2013.00233.

Bang, Woo-Dae, et al. "Pulmonary Hypertension Associated with Use of Phentermine." *Yonsei Medical Journal* 51.6 (2010): 971. https://doi.org/10.3349/ymj.2010.51.6.971.

Bay NVC. "List of Needs." BayNVC. https://baynvc.org/list-of-needs/.

BBB Australia. "Meet Your Guard Dog and Your Wise Owl." buildingbetterbrains.com.au, Dec. 25, 2019. https://buildingbetterbrains.com.au/guard-dog-and-wise-owl/.

Beck, Don, Teddy Hebo Larsen, Sergey Solonin, Rica Viljoen, Thomas Q. Johns. *Spiral Dynamics in Action*. London: Wiley, 2018. Kindle.

Benedetti, Fabrizio, et al. "How Placebos Change the Patient's Brain." *Neuropsychopharmacology* 36.1: 339–54. Jun. 30, 2010. https://doi.org/10.1038/npp.2010.81.

Benowitz, Neal L. "Pharmacology of Nicotine: Addiction, Smoking-Induced Disease, and Therapeutics." *Annual Review of Pharmacology and Toxicology* 49.1 (Sep. 27, 2010) 57–71. https://doi.org/10.1146/annurev.pharmtox.48.113006.094742.

Berger, Michele W. "Does More Money Correlate with Greater Happiness?" Penn Today, Mar. 6, 2023. https://penntoday.upenn.edu/news/does-more-money-correlate-greater-happiness-Penn-Princeton-research.

Biounity. "Dimenhydrinate." https://www.bionity.com/en/encyclopedia/Dimenhydrinate.html.

Brand, Russel. *Recovery: Freedom from Our Addictions*. New York: Picador, 2017.

Brean, Joseph. "The Lottery Curse: How the Unfathomable Good Fortune of Winning $50M Can Go Terribly Wrong." *National Post*, May 22, 2020. https://nationalpost.com/news/the-lottery-curse-how-the-unfathomable-good-fortune-of-winning-50m-can-go-terribly-wrong.

Britannica. "Seven Deadly Sins" *Encyclopedia Britannica*, Sep. 29, 2025. https://www.britannica.com/topic/seven-deadly-sins.

Brown, Brené. *Daring Greatly*. 1st ed. New York: Avery, 2012. Kindle.

Burritt, Dan. "Hardy and Amelia Leighton's Death due to Fentanyl, Other Drugs, Coroner Says." CBC News, Jul. 29, 2015. https://www.cbc.ca/news/canada/british-columbia/hardy-and-amelia-leighton-s-death-due-to-fentanyl-other-drugs-coroner-says-1.3173286.

Canna, Sarah, and St. Clair, Carley. "Neuroscience Insights on Radicalization and Mobilization to Violence: A Review." Dec. 14, 2012. http://nsiteam.com/social/wp-content/uploads/2016/01/Neuroscience-Insights-on-%20Radicalization-and-Mobilization-to-Violence-November-2012.pdf.

Carl, J., et al. "Intravenous Infusion of Nitroglycerine Leads to Increased Permeability on DynamicContrast-Enhanced MR Imaging in Pig Brains." *American Journal of Neuroradiology*, 36.7 (July 2015) 1288–1292. https://doi.org/10.3174/ajnr.a4279

Carney, Mark. *Value(s) : Building a Better World for All*. Toronto: McClelland & Stewart, 2021. Kindle.

Castillo, Michelle. "Can Alcohol Make Men Smarter? Study Suggests Yes." CBS News, Apr. 12, 2012. https://www.cbsnews.com/news/can-alcohol-make-men-smarter-study-suggests-yes/.

CBS News. "Tears for Fears returns with 'The Tipping Point.'" Feb. 20, 2022. https://www.cbsnews.com/news/tears-for-fears-the-tipping-point/.

CCSA. "Canada's Guidance on Alcohol and Health." Canadian Centre on Substance Use and Addiction, Jan. 4, 2023. https://www.ccsa.ca/canadas-guidance-alcohol-and-health.

Cesura, Andrea. "Monoamine Oxidase - an Overview." Science Direct, 2007. https://www.sciencedirect.com/topics/biochemistry-genetics-and-molecular-biology/monoamine-oxidase.

CFI Team. "Neoclassical Economics." Corporate Finance Institute. https://corporatefinanceinstitute.com/resources/economics/neoclassical-economics/.
Chong, Pearlynne L. H., et al. "Sleep, Cerebrospinal Fluid, and the Glymphatic System: A Systematic Review." *Sleep Medicine Reviews* 61 (Feb. 1, 2022). https://doi.org/10.1016/j.smrv.2021.101572.
Cleveland Clinic. "Epinephrine (Adrenaline): What It Is, Function, Deficiency & Side Effects." Mar. 27, 2022. https://my.clevelandclinic.org/health/articles/22611-epinephrine-adrenaline.
Cloud, John. "The Lure of Ecstacy." *Time Magazine*, Jun. 5, 2000. https://content.time.com/time/subscriber/article/0,33009,997083,00.html.
CNET. "PayPal to Levy Fines for Gambling, Porn." Oct. 10, 2004. https://www.cnet.com/culture/paypal-to-levy-fines-for-gambling-porn/.
Collins, Francis. "Imaging Willpower: Using Brain Scans to Explore Obesity." NIH Director's Blog, Nov. 25, 2014. https://directorsblog.nih.gov/2014/11/25/using-brain-scans-to-explore-obesity/.
Collins, Sonya. "What You Need to Know about Ketamine's Effects." WebMD, Apr. 14, 2015. https://www.webmd.com/depression/features/what-does-ketamine-do-your-brain.
Confessore, Nicholas. "Cambridge Analytica and Facebook: The Scandal and the Fallout so Far." *The New York Times*, Apr. 4, 2018. https://www.nytimes.com/2018/04/04/us/politics/cambridge-analytica-scandal-fallout.html.
Cox, Brittney M., et al. "Oxytocin Acts in Nucleus Accumbens to Attenuate Methamphetamine Seeking and Demand." *Biological Psychiatry* 81.11 (Jun. 2017) 949–58. https://www.ncbi.nlm.nih.gov/pmc/articles/PMC5432412/.
Cuncic, Arlin. "What Is Emotional Lability?" Verywell Mind, Sep. 26, 2021. https://www.verywellmind.com/what-is-emotional-lability-5200864.
Darling, Monique. *Beyond Cuddle Party*. Orem, UT: Juicy Enlightenment, 2016.
David, Sharoon, and Paras B. Khandhar. "Double-Blind Study." PubMed. Treasure Island, FL: StatPearls, 2022. https://www.ncbi.nlm.nih.gov/books/NBK546641/.
Dawkins, Richard. *The Selfish Gene*. Oxford University Press, 2016.
De Waal, Frans B.M. "What Is an Animal Emotion?" *Annals of the New York Academy of Sciences* 1224.1 (Apr. 2011) 191–206. https://doi.org/10.1111/j.1749-6632.2010.05912.x.
———. "Two Monkeys Were Paid Unequally: Excerpt from Frans de Waal's TED Talk." YouTube, 2014. https://youtu.be/meiU6TxysCg.
DeAngelis, Tori. "Are Beliefs Inherited?" apa.org, Apr. 2004. https://www.apa.org/monitor/apr04/beliefs.
DeYoung, Kevin. "Two Cheers for the Spirituality of the Church." The Gospel Coalition, Jan. 31, 2019. https://www.thegospelcoalition.org/blogs/kevin-deyoung/two-cheers-spirituality-church/.
Dholakia, Nazish. "Fifty Years Ago Today, President Nixon Declared the War on Drugs." Vera Institute of Justice, Jun. 17, 2021. https://www.vera.org/news/fifty-years-ago-today-president-nixon-declared-the-war-on-drugs.
Di Pietro, Paola. "Sensing God: Learning to Meditate during Lent." WCCM, Feb. 17, 2021. https://wccm.org/news/sensing-god-learning-to-meditate-during-lent/.
Dingman, Mark. *Bizarre*. London: John Murray One, 2024. Kindle.

Dion, Lisa. "Are Guilt, Shame, and Worry Rewarding? You Might Be Surprised." Synergetic Play Therapy Institute, Oct. 17, 2017. https://synergeticplaytherapy.com/are-guilt-shame-and-worry-rewarding-you-might-be-surprised/.

Duffy, James D. "A Primer on Integral Theory and Its Application to Mental Health Care." *Global Advances in Health and Medicine* 9.1 (Sep. 21, 2020). https://doi.org/10.1177/2164956120952733.

Dupuy, John. *Integral Recovery : A Revolutionary Approach to the Treatment of Alcoholism and Addiction.* Albany, NY: Excelsior Editions/State University Of New York Press, 2013. Kindle.

Duran, Domingos, et al. "Guidelines for the Intervention in Dissuasion." sicad.pt, 2013. https://sicad.pt/BK/Intervencao/Dissuasao/Documents/LOID_EN.pdf.

Earth Balance. "Parasympathetic System and Tai Chi." Earth Balance Tai Chi, Dec 11, 2010. https://earthbalance-taichi.com/2010/12/parasympathetic-system-and-tai-chi/.

Eckersley, Nicole. "The Eyes Have It: How Staring at Strangers Became a Global Movement." *The Guardian*, Jan. 10, 2017. https://www.theguardian.com/culture/2017/jan/10/the-eyes-have-it-how-staring-at-strangers-became-a-global-movement.

Edinoff, Amber N., et al. "Oxytocin, a Novel Treatment for Methamphetamine Use Disorder." *Neurology International* 14.1 (Jan. 30, 2022) 186–98. https://doi.org/10.3390/neurolint14010015.

Edwards, Jim. "Yes, Bayer Promoted Heroin for Children." Business Insider, Nov. 2011. https://www.businessinsider.com/yes-bayer-promoted-heroin-for-children-here-are-the-ads-that-prove-it-2011-11.

Erowid. "Erowid DXM (Dextromethorphan, DM) Vault." https://www.erowid.org/chemicals/dxm/dxm.shtml.

Favini, John. "What If Competition Isn't as 'Natural' as We Think?" *Slate Magazine*, Jan. 23, 2020. https://slate.com/technology/2020/01/darwin-competition-collaboration-evolutionary-biology-climate-change.html.

Forbes Magazine. "Inside the Dark Hidden World of the Fentanyl Industry." Nov. 6, 2019. https://www.forbes.com/sites/quora/2019/11/06/inside-the-dark-hidden-world-of-the-fentanyl-industry/

———. "The Wages of Sin." Apr. 25, 2005. https://www.forbes.com/forbes/2005/0425/046.html?sh=5d9f14177830.

Ford, Debbie. *The Dark Side of the Light Chasers.* New York: Penguin, 2010. Kindle.

Frankl, Victor. *Man's Search for Meaning.* Boston: Beacon, 200. Kindle

———. *The Unconscious God.* New York: Pocket, 1985.

———. *The Unheard Cry for Meaning.* New York: Touchstone, 2011.

Free Hugs Campaign. "Free Hugs Campaign." 2019. https://www.freehugscampaign.org/.

Freeman, Ellen W. "Premenstrual Dysphoric Disorder: Recognition and Treatment." Psychiatrist.com, Feb. 2003. https://www.psychiatrist.com/pcc/premenstrual-dysphoric-disorder-recognition-treatment/.

Fries, Eva, et al. "The Cortisol Awakening Response (CAR): Facts and Future Directions." *International Journal of Psychophysiology* 72.1 (Apr. 2009) 67–73. https://doi.org/10.1016/j.ijpsycho.2008.03.014.

FritoLay. "Product Locator." https://contact.pepsico.com/fritolay/product-locator.

Ganz, Brandi. "Unpacking and Dispelling the Concept of Rock-Bottom." Recovery Unplugged, Jan. 16, 2017. https://www.recoveryunplugged.com/unpacking-dispelling-concept-rock-bottom/.

Gasperi, Marianna, et al. "Pain and Trauma: The Role of Criterion a Trauma and Stressful Life Events in the Pain and PTSD Relationship." *The Journal of Pain* 22.11 (Nov. 10, 2021) 1506–17. https://www.ncbi.nlm.nih.gov/pmc/articles/PMC8578317/.

Gass, J. T., and L. J. Chandler. "The Plasticity of Extinction: Contribution of the Prefrontal Cortex in Treating Addiction through Inhibitory Learning." *Frontiers in Psychiatry* 4.46 (2013). https://doi.org/10.3389/fpsyt.2013.00046.

Germa, Fikre. "Stethoscopes and Stories." *Canadian Family Physician* 63.8 (Aug. 1, 2017) 626–27. https://www.ncbi.nlm.nih.gov/pmc/articles/PMC5555332/.

Ghosh, Karthik, et al. "Mindfulness Using a Wearable Brain Sensing Device for Health Care Professionals during a Pandemic: A Pilot Program." *Journal of Primary Care & Community Health* 14: 21501319231162308 (Mar. 23, 2023). https://doi.org/10.1177/21501319231162308.

Gjedde, Albert, et al. "Inverted-U-Shaped Correlation between Dopamine Receptor Availability in Striatum and Sensation Seeking." *Proceedings of the National Academy of Sciences* 107.8 (Feb. 4, 2010) 3870–75. https://doi.org/10.1073/pnas.0912319107.

Gleick, James. "SURVIVAL of the LUCKIEST." *The New York Times*, Oct. 22, 1989. https://www.nytimes.com/1989/10/22/books/survival-of-the-luckiest.html.

Goeders, Nick. "Stress and Cocaine Addiction." JPET, Jun. 2002. https://jpet.aspetjournals.org/article/S0022-3565(24)35674-5/abstract.

Gometz, Emma. "The Hidden Physics In Van Gogh's 'The Starry Night.'" Science Friday. Sep. 27, 2024. https://www.sciencefriday.com/segments/physics-van-gogh-starry-night/.

Graf, Evan. N., et al. "Corticosterone Acts in the Nucleus Accumbens to Enhance Dopamine Signaling and Potentiate Reinstatement of Cocaine Seeking." *Journal of Neuroscience* 33.29 (Jul. 17, 2013) 11800–11810. https://doi.org/10.1523/jneurosci.1969-13.2013.

Gray, Elizabeth Kelly. "American Opium Dens, 1850–1910." *Oxford University Press EBooks*, 147-C7.F3. Sep. 2, 2023, https://doi.org/10.1093/oso/9780190073121.003.0008.

Gregoire, Carolyn. "MDMA Changes the Way People Talk about Their Loved Ones: Study." *HuffPost*, Apr. 30, 2015. https://www.huffpost.com/entry/mdma-therapy_n_7181200.

Gregory, T. Ryan. "Understanding Natural Selection: Essential Concepts and Common Misconceptions." *Evolution: Education and Outreach* 2.2 (Apr. 9, 2009) 156–75. https://doi.org/10.1007/s12052-009-0128-1.

Griffiths, Mark, and M. M. Auer. "Becoming Hooked? Angling, Gambling, and 'Fishing Addiction.'" https://core.ac.uk/download/pdf/196213906.pdf.

Grisel, Judith. *Never Enough: The Neuroscience and Experience of Addiction*. New York: Doubleday, 2019. Kindle.

Grof, Stanislav. "Beyond Psychoanalysis - the Future of Psychiatry: Conceptual Challenges to Psychiatry, Psychology, and Psychotherapy by Stanislav Grof M.D." www.newdualism.org. https://www.newdualism.org/papers/S.Grof/sgrof.htm.

———. *Realms of the Human Unconscious : Observations from LSD Research*. London: Souvenir, 2019.

Gunter, Melissa. "6 Plants That Contain Caffeine (with Pictures)." Coffee Affection, Mar. 29, 2022. https://coffeeaffection.com/plants-that-contain-caffeine/.

Halber, Deborah. "Motivation: Why You Do the Things You Do." Brainfacts.org, 2018. https://www.brainfacts.org/Thinking-Sensing-and-Behaving/Learning-and-Memory/2018/Motivation-Why-You-Do-the-Things-You-Do-082818.

Hall-Flavin, Daniel K. "Avoid the Combination of High-Tyramine Foods and MAOIs." Mayo Clinic, Dec. 18, 2018. https://www.mayoclinic.org/diseases-conditions/depression/expert-answers/maois/faq-20058035.

Hall, Wayne, and Megan Weier. "Lee Robins' Studies of Heroin Use among US Vietnam Veterans." *Addiction* 112.1 (Sep. 20, 2016) 176–80. https://doi.org/10.1111/add.13584.

Hartney, Elizabeth. "The Comedown, Crash, or Rebound Effect You Get after Taking Drugs." Verywell Mind, Apr. 12, 2023. https://www.verywellmind.com/comedown-crash-rebound-effect-after-drugs-4171269.

Hawking, Tom. "The Collected Wisdom of David Bowie." Flavorwire, Mar. 15, 2013. https://www.flavorwire.com/377621/the-collected-wisdom-of-david-bowie.

Hayes, Jack. "The Opium Wars in China." asiapacificcurriculum.ca, 2023. https://asiapacificcurriculum.ca/learning-module/opium-wars-china.

Health Canada. "Low-risk alcohol drinking guidelines." Jul. 5, 2021. https://www.canada.ca/en/health-canada/services/substance-use/alcohol/low-risk-alcohol-drinking-guidelines.html.

Herbig, Britta, and Andreas Glöckner. "Experts and Decision Making: First Steps towards a Unifying Theory of Decision Making in Novices, Intermediates and Experts." *SSRN Electronic Journal* (Feb. 4, 2009). https://doi.org/10.2139/ssrn.1337449.

Heshmat, Shahram. "The Role of Denial in Addiction." Psychology Today, Nov. 13, 2018. https://www.psychologytoday.com/ca/blog/science-choice/201811/the-role-denial-in-addiction.

Holtzhausen, Leon. "Addiction – a Brain Disorder or a Spiritual Disorder." Oatext, https://www.oatext.com/Addiction-a-brain-disorder-or-a-spiritual-disorder.php.

Holwerda, Edwin. "Home · Spiral Dynamics Integral." Spiral Dynamics Integral. https://spiraldynamicsintegral.nl/en/.

House, Patrick. *Nineteen Ways of Looking at Consciousness*. New York: St. Martin's. 2022. Kindle Edition.

Howard, Kimberly, et al. "Early Mother–Child Separation, Parenting, and Child Well-Being in Early Head Start Families." *Attachment & Human Development* 13.1 (Jan. 13, 2011) 5–26. https://doi.org/10.1080/14616734.2010.488119.

Howes, Oliver D., and Shitji Kapur. "The Dopamine Hypothesis of Schizophrenia: Version III—the Final Common Pathway." *Schizophrenia Bulletin* 35.3 (Mar. 30, 2009) 549–62. https://doi.org/10.1093/schbul/sbp006.

Hoy, Chris. "Who Is Sir Chris Hoy?" Mar. 27, 2019. https://chrishoy.co.uk/profile/.

Humans of New York. Medellín, Colombia, Apr. 20, 2017. https://www.facebook.com/share/1FxC9bSUAs/.

Husserl, Edward. *Ideas: General Introduction to Pure Phenomenology*. Eastford, CT: Martino, 1931.

Hyshka, Elaine, et al. "Harm Reduction in Name, but Not Substance: A Comparative Analysis of Current Canadian Provincial and Territorial Policy Frameworks." *Harm Reduction Journal* 14.1 (Jul. 26, 2017). https://doi.org/10.1186/s12954-017-0177-7.

IES Brain Research. "History." https://www.iesbrainresearch.org/history.
Isaacson, Walter. *Einstein: His Life and Universe*. New York: Simon and Schuster, 2007. eBook.
Israel, Salomon, et al. "Oxytocin Decreases Accuracy in the Perception of Social Deception." *Psychological Science* 25.1 (Nov. 13, 2013) 293–95. https://doi.org/10.1177/0956797613500794.
Jenkins, Bill. "Why Your Brain Loves Chocolate." Scientific Learning, Feb. 10, 2011. https://www.scilearn.com/why-your-brain-loves-chocolate/.
Jiang, Rena. "Two Sides of the Same Coin: MSG and Umami." Wu Tsai Neurosciences Institute, Apr. 24, 2017. https://www.neuwritewest.org/blog/two-sides-of-the-same-coin-msg-and-umami.
Juber, Mahamad. "Health Benefits of Capsaicin." WebMD, Nov. 29, 2022. https://www.webmd.com/diet/health-benefits-capsaicin.
Jung, Carl. *Collected Works*. Vol. 9ii. Princeton University Press, 2023.
———. *Letters of C. G. Jung*. London: Routledge, 2015.
Kamei, Junzo, et al. "Antitussive Effect of β-Endorphin Is Mediated by μ-Opioid Receptors, but Not by κ- or ε-Opioid Receptors." *European Journal of Pharmacology* 233.2-3 (Mar. 1993) 251–54. https://doi.org/10.1016/0014-2999(93)90057-0.
Kaufman, Scott Barry. "What Does It Mean to Be Self-Actualized in the 21st Century?" Scientific American Blog Network, Nov. 7, 2018. https://blogs.scientificamerican.com/beautiful-minds/what-does-it-mean-to-be-self-actualized-in-the-21st-century/.
Kennedy, Vera. *Beyond Race: Cultural Influences on Human Social Life*. Davis, CA: LibreTexts, 2018.
Kerr, Fiona, et al. "Neurophysiology of Human Touch and Eye Gaze in Therapeutic Relationships and Healing." *JBI Database of Systematic Reviews and Implementation Reports* 17.2 (Feb. 2019) 209–47. https://doi.org/10.11124/jbisrir-2017-003549.
King, Stephen. *On Writing*. Scribner: New York, 2010.
Kotler, Steven. "Addicted to Bang: The Neuroscience of the Gun." Forbes, Dec. 18, 2012. https://www.forbes.com/sites/stevenkotler/2012/12/18/addicted-to-bang-the-neuroscience-of-the-gun.
Kühn, S, et al. "The Neural Basis of Video Gaming." *Translational Psychiatry* 1.11 (Nov. 15, 2011) e53–53. https://doi.org/10.1038/tp.2011.53.
Kurihara, Kenzo. "Umami the Fifth Basic Taste: History of Studies on Receptor Mechanisms and Role as a Food Flavor." *BioMed Research International* 2015. Volume 2015 (2015) 1–10. https://doi.org/10.1155/2015/189402.
Küstenmacher, Marion, et al. *God 9.0 (English)*. WIRmachenDRUCK GmbH, Backnang, 2017.
Kwong, Matt. "In a Land of Workaholics, Burned-out South Koreans Go to 'Prison' to Relax." CBC, Feb. 14, 2018. https://www.cbc.ca/news/world/south-korea-overwork-culture-jail-retreat-prison-inside-me-1.4527832.
Lagace, Martha. "The Simple Economics of Open Source." HBS Working Knowledge, May 16, 2000. https://hbswk.hbs.edu/item/the-simple-economics-of-open-source.
Le Page, Michael. "Chimps Are Not as Superhumanly Strong as We Thought They Were." New Scientist, Jun. 26, 2017. https://www.newscientist.com/article/2138714-chimps-are-not-as-superhumanly-strong-as-we-thought-they-were/.

Linden, David. "Video Games Can Activate the Brain's Pleasure Circuits" Psychology Today, Oct. 25, 2011. https://www.psychologytoday.com/ca/blog/the-compass-pleasure/201110/video-games-can-activate-the-brains-pleasure-circuits-0.

Lindsay, Bethany. "Atheist nurse wins fight to end mandatory 12-step addiction treatment for health staff in Vancouver." CBC, Dec. 11, 2019. https://www.cbc.ca/news/canada/british-columbia/bc-byron-wood-nurse-12-step-religious-discrimination-settlement-1.5391650

Living Art. "Living Art Productions - Rave: Articles." 2025. http://www.livingart.com/raving/articles/article11.htm.

Locklear, Mallory. "How Tainted Drugs 'Froze' Young People—but Kickstarted Parkinson's Research." Ars Technica, May 18, 2016. https://arstechnica.com/science/2016/05/medical-mystery-how-tainted-drugs-froze-young-people-but-kickstarted-parkinsons-research/.

Love, Tiffany M. "Oxytocin, Motivation and the Role of Dopamine." *Pharmacology, Biochemistry, and Behavior* 0 (Apr. 1, 2014) 49–60. https://doi.org/10.1016/j.pbb.2013.06.011.

Lulu, Nigistry. "Prescription Methamphetamine: Everything You Need to Know." WebMD. https://www.webmd.com/connect-to-care/addiction-treatment-recovery/methamphetamine/what-you-need-to-know-about-prescription-meth.

Lustig, Robert H. *The Hacking of the American Mind : The Science behind the Corporate Takeover of Our Bodies and Brains*. New York: Avery, 2017. Kindle.

Lyell, Charles. "Dopamine Profile: Why Power/Money/Esteem Addicts Are More Dangerous than Junkies." TheDopamineProject.org, Jan. 19, 2013. https://dopamineproject.org/2013/01/why-power-money-and-esteem-addicts-are-more-dangerous-than-junkies/.

———. "Dopamine Withdrawal." TheDopamineProject.org, Jul. 4, 2011. https://dopamineproject.org/dopamine-withdrawal/.

Ma, Yinxiang, et al. "Hidden turbulence in van Gogh's *The Starry Night*." *Physics of Fluids*, 36.9. (2024). https://doi.org/10.1063/5.0213627.

Mack Whitaker-Asmitia, Patricia. "The Discovery of Serotonin and Its Role in Neuroscience." nature.com, Aug. 1, 1999. https://www.nature.com/articles/1395355.

Margoczi, Gyozo. *Feelings - The Need for a new science*. Self-Published, 2016. Kindle.

Marinelli, Silvia, et al. "Activation of TRPV1 in the VTA Excites Dopaminergic Neurons and Increases Chemical and Noxious-Induced Dopamine Release in the Nucleus Accumbens." *Neuropsychopharmacology* 30.5 (Nov. 24, 2004) 864–70. https://doi.org/10.1038/sj.npp.1300615.

Marx, Karl, and Friedrich Engels. *Economic and Philosophic Manuscripts of 1844*. Buffalo, NY: Prometheus, 1988.

Marx, Karl. "A Contribution to the Critique of Hegel's Philosophy of Right 1844." www.marxists.org, 1844. https://www.marxists.org/archive/marx/works/1843/critique-hpr/intro.htm.

———. "Capital Vol. I - Chapter One." www.marxists.org, 1867. https://www.marxists.org/archive/marx/works/1867-c1/ch01.htm#S4.

———. "Capital Vol. I - Chapter Four." www.marxists.org, 1867. https://www.marxists.org/archive/marx/works/1867-c1/ch04.htm.

———. "Capital Vol. I - Chapter Twenty-Four." www.marxists.org, 1867. https://www.marxists.org/archive/marx/works/1867-c1/ch24.htm.

Maslow, Abraham. *Religions, Values, and Peak-Experiences*. Victoria, BC: Rare Treasures, 2021. Kindle.
Maté, Gabor, and Daniel Maté. *The Myth of Normal: Trauma, Illness, and Healing in a Toxic Culture*. New York: Penguin, 2022. Kindle.
Maté, Gabor. *In the Realm of Hungry Ghosts*. Toronto: Vintage Canada, 2008. Kindle.
Matthews, J. Scott, et al. "The Mediated Horserace: Campaign Polls and Poll Reporting." *Canadian Journal of Political Science / Revue Canadienne de Science Politique* 45.2 (2012) 261–87. https://www.jstor.org/stable/23320971.
Mayo Clinic. "Monoamine Oxidase Inhibitors (MAOIs)." 2016. https://www.mayoclinic.org/diseases-conditions/depression/in-depth/maois/art-20043992.
McDermott, Rose, et al. "Monoamine Oxidase a Gene (MAOA) Predicts Behavioral Aggression Following Provocation." *Proceedings of the National Academy of Sciences* 106.7 (Jan 23, 2009) 2118–23. https://doi.org/10.1073/pnas.0808376106.
McLeod, Saul. "Maslow's Hierarchy of Needs." Simply Psychology, Jul. 26, 2023. https://www.simplypsychology.org/maslow.html.
Medicines.org. "Multi-Action ACTIFED Tablets - Summary of Product Characteristics." https://www.medicines.org.uk/emc/product/6191/smpc.
Menahem, Mor. *The Second Jewish Revolt : The Bar Kokhba War, 132-136 CE*. Leiden: Brill, 2016.
Miller, Gregory M. "The Emerging Role of Trace Amine-Associated Receptor 1 in the Functional Regulation of Monoamine Transporters and Dopaminergic Activity." *Journal of Neurochemistry* 116.2 (Dec. 16, 2010) 164–76. https://doi.org/10.1111/j.1471-4159.2010.07109.x.
Miller, Kelly. "What Is Delayed Gratification? 5 Examples & Definition." Positive Psychology, Dec. 30, 2019. https://positivepsychology.com/delayed-gratification.
Minnesota DOH. "Opioids Prescribing Practices Perception of Pain." Oct. 3, 2022. https://www.health.state.mn.us/communities/opioids/prevention/painperception.html.
Murray, Krystina. "Drug Abuse Peaks after 50 Year War on Drugs." Addiction Center, Jul. 7, 2021. https://www.addictioncenter.com/news/2021/07/drug-use-peaks-after-50-year-war-on-drugs/.
Nader, Michael, et al. "Effects of Cocaine Self-Administration on Striatal Dopamine Systems in Rhesus Monkeys Initial and Chronic Exposure." *Neuropsychopharmacology* 27.1 (Jul. 2002) 35–46. https://doi.org/10.1016/s0893-133x(01)00427-4
National Institute on Drug Abuse. "What Are MDMA's Effects on the Brain?" 2017. https://nida.nih.gov/publications/research-reports/mdma-ecstasy-abuse/what-are-mdmas-effects-on-brain.
Newberg, Andrew, and Waldman, Mark. *How God Changes Your Brain*. New York: Ballantine, 2009. Kindle.
Nguyen, Linda, et al. "Involvement of Sigma-1 Receptors in the Antidepressant-like Effects of Dextromethorphan." Edited by Allan Siegel. *PLoS ONE* 9.2 (Feb. 28, 2014) e89985. https://doi.org/10.1371/journal.pone.0089985.
NHRC. "Principles of Harm Reduction." National Harm Reduction Coalition. https://harmreduction.org/about-us/principles-of-harm-reduction/.
NIDA. "Genetics and Epigenetics of Addiction DrugFacts." National Institute on Drug Abuse, Aug. 5, 2019. https://nida.nih.gov/publications/drugfacts/genetics-epigenetics-addiction.
O'Leary, Daniel. *Escaping the Progress Trap*. Montreal: Geozone Communications, 2007.

Ostrin, Lisa A., and Adrian Glasser. "The Effects of Phenylephrine on Pupil Diameter and Accommodation in Rhesus Monkeys." *Investigative Ophthalmology & Visual Science* 45.1 (Jan. 1, 2004) 215–21. https://www.ncbi.nlm.nih.gov/pmc/articles/PMC2913435/.

Palm, Sara, and Ingrid Nylander. "Endorphin - an Overview." Science Direct, 2016. https://www.sciencedirect.com/topics/veterinary-science-and-veterinary-medicine/endorphin.

Panarchy.org. "Arthur Koestler, Some General Properties of Self-Regulating Open Hierarchic Order (1969)." https://panarchy.org/koestler/holon.1969.html.

Pandey, Sanjay, and Prachaya Srivanitchapoom. "Levodopa-Induced Dyskinesia: Clinical Features, Pathophysiology, and Medical Management." *Annals of Indian Academy of Neurology* 20.3 (2017) 190–98. https://doi.org/10.4103/aian.AIAN_239_17.

Pangambam, S. "Everything You Think You Know about Addiction Is Wrong by Johann Hari (Full Transcript)." The Singju Post, Jul. 22, 2015. https://singjupost.com/everything-you-think-you-know-about-addiction-is-wrong-by-johann-hari-full-transcript/?singlepage=1.

———. "Gary Wilson Discusses the Great Porn Experiment (Transcript)." The Singju Post, Jun. 19, 2014. https://singjupost.com/gary-wilson-discusses-great-porn-experiment-transcript/.

———. "The Power of Vulnerability by Brene Brown (Transcript)." The Singju Post, Sep. 5, 2014. https://singjupost.com/power-vulnerability-brene-brown-transcript/?singlepage=1.

Parker, Matt. "Win a million dollars with maths, No. 3: The Navier-Stokes equations." *The Guardian*, Dec. 15, 2010. https://www.theguardian.com/science/blog/2010/dec/14/million-dollars-maths-navier-stokes.

Peele, Stanton, and Archy Brodsky. *Love and Addiction*. New York: Taplinger, 1975. Kindle.

Peters, Steve. *The Chimp Paradox : The Mind Management Program to Help You Achieve Success, Confidence, and Happiness*. New York: Jeremy P. Tarcher/Penguin, 2013. Kindle.

Piper, John. *Desiring God: Meditations of a Christian Hedonist*. Colorado Springs: Multnomah, 2017. Kindle.

Pollack, Robert. *The Faith of Biology & the Biology of Faith*. Columbia University Press, 2013. Kindle.

Prah, Alja, et al. "Brunner Syndrome Caused by Point Mutation Explained by Multiscale Simulation of Enzyme Reaction." *Scientific Reports* 12.1 (Dec. 19, 2022) 21889. https://doi.org/10.1038/s41598-022-26296-7.

Presley, Priscilla. *Elvis and Me*. Blackstone, Ashland, OR, 2022. Kindle.

Provincial Health Officer. "Provincial Health Officer Declares Public Health Emergency." news.gov.bc.ca, Apr. 14, 2016. https://news.gov.bc.ca/releases/2016HLTH0026-000568.

PubChem. "Carfentanil." https://pubchem.ncbi.nlm.nih.gov/compound/Carfentanil.

Public Safety and Solicitor General. "More than 1,600 Lives Lost to Illicit Drugs in First Nine Months of 2022." BC Government, Nov. 7, 2022. https://news.gov.bc.ca/releases/2022PSSG0069-001656.

Purves-Tyson, Tertia D., Samantha J. Owens, Kay L. Double, Reena Desai, David J. Handelsman, and Cynthia Shannon Weickert. "Testosterone Induces Molecular Changes in Dopamine Signaling Pathway Molecules in the Adolescent Male Rat Nigrostriatal Pathway." *PLoS ONE* 9.3 (Mar. 11, 2014) e91151. https://doi.org/10.1371/journal.pone.0091151.

Quora, Contributor. "Why Are Younger People More Creative than Adults?" Slate Magazine, Aug. 6, 2016. https://slate.com/human-interest/2016/08/why-are-younger-people-more-creative-than-adults.html.

Race, Paul. "A Brief History of Contemporary Christian Music." schooloftherock.com, 2016. https://schooloftherock.com/html/a_brief_history_of_contemporar.html.

Ro.co. "Amitriptyline: Everything You Need to Know." Health Guide. https://ro.co/health-guide/amitriptyline-101/.

Rock, David. "The Neuroscience of Mindfulness." Psychology Today, Oct. 11, 2009. https://www.psychologytoday.com/ca/blog/your-brain-at-work/200910/the-neuroscience-of-mindfulness.

Rodríguez de Fonseca, F., and M. Navarro. "Role of the Limbic System in Dependence on Drugs." *Annals of Medicine* 30.4 (Aug. 1, 1998) 397–405. https://doi.org/10.3109/07853899809029940.

Ronningstam, Elsa, and Arielle R. Baskin-Sommers. "Fear and Decision-Making in Narcissistic Personality Disorder—a Link between Psychoanalysis and Neuroscience." *Dialogues in Clinical Neuroscience* 15.2 (Jun. 1, 2013) 191–201. https://www.ncbi.nlm.nih.gov/pmc/articles/PMC3811090/.

Rosenberg, Marshall. *Non-Violent Communication: A Language of Life.* Encinitas, CA: Puddledancer, 2015. Kindle.

Rosenthal, Talma. "Device-Guided Breathing Exercises Reduce Blood Pressure: Ambulatory and Home Measurements." *American Journal of Hypertension* 14.1 (Jan. 2001) 74–76. https://doi.org/10.1016/s0895-7061(00)01235-8.

Roth, Bryan L. "Molecular Pharmacology of Metabotropic Receptors Targeted by Neuropsychiatric Drugs." *Nature Structural & Molecular Biology* 26.7 (Jul. 2019) 535–44. https://doi.org/10.1038/s41594-019-0252-8.

Rubarth, Scott. "Stoic Philosophy of Mind." Internet Encyclopedia of Philosophy. https://iep.utm.edu/stoicmind/.

Runsewe, Damilola. "The Inhibitory Effects of Tetracycline on Mastocytosis." *Journal of Biosciences and Medicines* 7.6 (2019) 33–58. https://doi.org/10.4236/jbm.2019.76004.

Sacks, Oliver. "A Bolt from the Blue." *The New Yorker*, Jul. 16, 2007. https://www.newyorker.com/magazine/2007/07/23/a-bolt-from-the-blue.

———. *Musicophilia*. New York: Vintage, 2008.

Sandberg, Erica. "Antidepressants and Compulsive Shopping." RxISK, Dec. 4, 2012. https://rxisk.org/rxisk-stories-antidepressants-and-compulsive-shopping/.

Sapolsky, Robert. *Dopamine Jackpot! Sapolsky on the Science of Pleasure.* YouTube, 2012. https://youtu.be/axrywDP9Iio.

Scharmer, Otto. *The Essentials of Theory U: Core Principles and Applications.* Oakland, CA: Berrett-Koehler, 2018.

Schieber, Jonathan. "Meet the Tech Company That Wants to Make You Even More Addicted to Your Phone." TechCrunch, Sep. 8, 2017. https://techcrunch.com/2017/09/08/meet-the-tech-company-that-wants-to-make-you-even-more-addicted-to-your-phone/.

Schultz, Wolfram. "Dopamine Reward Prediction Error Coding." *Dialogues in Clinical Neuroscience* 18.1 (Mar. 2016) 23–32. https://doi.org/10.31887/dcns.2016.18.1/wschultz.

Schwarcz, Joe. "The 'Chemical of Love.'" McGill University Office for Science and Society, Feb. 13, 2020. https://www.mcgill.ca/oss/article/nutrition/phenylethylamine-chemical-love.

Schwartz, Brian. "Billionaire Peter Thiel Gives First Six-Figure Donation of the Midterm Campaign Cycle to the RNC." CNBC, Aug. 29, 2018. https://www.cnbc.com/2018/08/29/peter-thiel-gives-six-figures-to-republican-national-committee.html.

Scott, Cameron. "Parkinson's Drugs May Lead to Compulsive Gambling, Shopping, And Sex" Healthline, Oct. 20, 2014. https://www.healthline.com/health-news/parkinsons-drugs-may-lead-to-compulsive-gambling-102014.

Sederer, Lloyd. "What Does 'Rat Park' Teach Us about Addiction?" *Psychiatric Times*, Jun. 10, 2019. https://www.psychiatrictimes.com/view/what-does-rat-park-teach-us-about-addiction.

Sher, Leo. "Cortisol and Seasonal Changes in Mood and Behavior." *Psychiatric Times*, 23. Oct. 1, 2006. https://www.psychiatrictimes.com/view/cortisol-and-seasonal-changes-mood-and-behavior.

Shulgin, Alexander T., and Ann Shulgin. *Pihkal: A Chemical Love Story*. Berkeley, CA: Transform, 2019.

———. *Tihkal: The Continuation*. Berkeley, CA: Transform, 2017.

Smith, Adam. *An Inquiry into the Nature and Causes of the Wealth of Nations*. Lawrence, KS: digireads.com, 2009.

Smith, Michael L., et al. "Methamphetamine and Amphetamine Isomer Concentrations in Human Urine Following Controlled Vicks VapoInhaler Administration." *Journal of Analytical Toxicology* 38.8 (October 1, 2014) 524–27. https://doi.org/10.1093/jat/bku077.

Stanley, Andy. "How to Get What You Really Want." Your Move, Jan. 13, 2020. https://yourmove.is/how-to-get-what-you-really-want-2/.

Statista. "Drug Overdose Deaths in Portugal 2008-2021." https://www.statista.com/statistics/911927/drug-overdose-deaths-in-portugal/.

———. "U.S. Wealth Distribution in 2016." 2024. https://www.statista.com/statistics/203961/wealth-distribution-for-the-us/.

Stebner, Beth. "How Your Phone Is Ruining Your Sex Life." StyleCaster, Nov. 11, 2015. https://stylecaster.com/lifestyle/love-sex/496772/how-smartphones-affect-your-sex-life/.

Stephens, Mary Ann C., and Gary Wand. "Stress and the HPA Axis: Role of Glucocorticoids in Alcohol Dependence." *Alcohol Research: Current Reviews* 34.4 (2012) 468–83. https://pubmed.ncbi.nlm.nih.gov/23584113/.

Sun, Lixing. *The Fairness Instinct*. Amherst, NY: Prometheus, 2013. Kindle.

Susana Peciña, Kent C. Berridge. "Hedonic Hot Spot in Nucleus Accumbens Shell: Where Do μ-Opioids Cause Increased Hedonic Impact of Sweetness?" PubMed Central. Dec. 14, 2005. https://pmc.ncbi.nlm.nih.gov/articles/PMC6726018/.

Swanson, Ana. "Study Finds Young Men Are Playing Video Games instead of Getting Jobs." *Chicago Tribune*, Sep. 23, 2016. https://www.chicagotribune.com/2016/09/23/study-finds-young-men-are-playing-video-games-instead-of-getting-jobs/.

Tennant, Forest. "Elvis Presley: Head Trauma, Autoimmunity, Pain, and Early Death." Practical Pain Management, 2013. https://www.practicalpainmanagement.com/pain/other/brain-injury/elvis-presley-head-trauma-autoimmunity-pain-early-death.

Tierney, Adrienne L, and Charles A Nelson. "Brain Development and the Role of Experience in the Early Years." *Zero to Three* 30.2 (2009) 9–13. https://www.ncbi.nlm.nih.gov/pmc/articles/PMC3722610/.

Tolle, Eckhart. *The Power of NOW : A Guide to Spiritual Enlightenment*. Novato, CA: Namaste, 2004.

Ulanov, Ann, and Barry Ulanov. *Religion and the Unconscious*. Philadelphia: Westminster, 1975.

University of Minnesota. "Love on the (Lion) Brain" cbs.umn.edu, Sep. 11, 2023. https://cbs.umn.edu/blog-posts/love-lion-brain.

Van Nueten, Jan M., et al. "Serotonin and Vascular Reactivity." *Pharmacological Research Communications* 17.7 (Jul. 1985) 585–608. https://doi.org/10.1016/0031-6989(85)90067-0.

Volkow, N D, et al. "Caffeine Increases Striatal Dopamine D2/D3 Receptor Availability in the Human Brain." *Translational Psychiatry* 5.4 (Apr. 2015) e549–49. https://doi.org/10.1038/tp.2015.46.

Walsh, Erin, and David Walsh. "When Phones Get in the Way of Connection." Psychology Today, Jan. 4, 2023. https://www.psychologytoday.com/ca/blog/smart-parenting-smarter-kids/202301/when-phones-get-in-the-way-of-connection.

Wardle, Margaret C., and Harriet de Wit. "MDMA Alters Emotional Processing and Facilitates Positive Social Interaction." *Psychopharmacology* 231.21 (Apr. 12, 2014) 4219–29. https://doi.org/10.1007/s00213-014-3570-x.

WebMD. "Methamphetamine (Desoxyn) – Uses, Side Effects, and More." Feb. 16, 2025. https://www.webmd.com/drugs/2/drug-9124/desoxyn-oral/details

Wehr, Gerhard. *Jung : A Biography*. Boston, MA; London: Shambhala, 2001.

West, Jean. "Children's Drug Is More Potent than Cocaine." *The Guardian*, Sep. 9, 2001. https://www.theguardian.com/uk/2001/sep/09/health.medicalscience.

Whitlock, Jennifer. "Discover When a Ventilator Is Necessary During and After Surgery." Verywell Health, Sep. 27, 2022. https://www.verywellhealth.com/when-a-ventilator-is-necessary-3156902.

Wikidoc. "Acacia Berlandieri." www.wikidoc.org. https://www.wikidoc.org/index.php/Acacia_berlandieri.

Wikipedia. "Amygdala." https://en.wikipedia.org/wiki/Amygdala.

———. "End-Plate Potential." https://en.wikipedia.org/wiki/End-plate_potential.

———. "End-Plate Potential." https://en.wikipedia.org/wiki/End-plate_potential.

———. "Parke-Davis." https://en.wikipedia.org/wiki/Parke-Davis.

———. "Positron Emission Tomography." https://en.wikipedia.org/wiki/Positron_emission_tomography.

———. "Purdue Pharma." https://en.wikipedia.org/wiki/Purdue_Pharma.

———. "Spirituality." https://en.wikipedia.org/wiki/Spirituality.

———. "Nefazodone." https://en.wikipedia.org/wiki/Nefazodone.

Wikiwand. "Fluoxetine." https://www.wikiwand.com/en/Fluoxetine.

Wilber, Ken. *A Brief History of Everything*. Boulder, CO: Shambhala, 1996. Kindle =.

———. *Integral Spirituality*. Boston: Shambhala, 2007. Kindle.

Wilcox, Bradford, and Samuel Sturgeon. "Too Much Netflix, Not Enough Chill: Why Young Americans Are Having Less Sex." *POLITICO Magazine*, Feb. 8, 2018. https://www.politico.com/magazine/story/2018/02/08/why-young-americans-having-less-sex-216953/.

Wittenberg, Ruthie E., et al. "Nicotinic Acetylcholine Receptors and Nicotine Addiction: A Brief Introduction." *Neuropharmacology* 177.108256 (Jul. 2020) 108256. https://doi.org/10.1016/j.neuropharm.2020.108256.

Wong, Julia. "Former Facebook Executive: Social Media Is Ripping Society Apart." *The Guardian*, Dec. 12, 2017. https://www.theguardian.com/technology/2017/dec/11/facebook-former-executive-ripping-society-apart.

Woodrow, Ralph. *The Babylon Connection?* Palm Springs, CA: Ralph Woodrow, 1997.

World Health Organization. "No level of alcohol consumption is safe for our health." Jan 4, 2023. https://www.who.int/europe/news/item/04-01-2023-no-level-of-alcohol-consumption-is-safe-for-our-health.

Yardley, William. "Debbie Ford, Author of Self-Help Books, Is Dead at 57." *The New York Times*, Feb. 21, 2013. https://www.nytimes.com/2013/02/21/books/debbie-ford-57-author-of-motivational-books.html.

Zaccaro, Andrea, et al. "How Breath-Control Can Change Your Life: A Systematic Review on Psycho-Physiological Correlates of Slow Breathing." *Frontiers in Human Neuroscience* 12.353 (Sep. 7, 2018) 1–16. https://doi.org/10.3389/fnhum.2018.00353.

Zalewska-Kaszubska, Jadwiga, and Elżbieta Czarnecka. "Deficit in Beta-Endorphin Peptide and Tendency to Alcohol Abuse." *Peptides* 26.4 (Apr. 2005) 701–5. https://doi.org/10.1016/j.peptides.2004.11.010.

Zepinic, Vito. "Disintegration of the Self-Structure Caused by Severe Trauma." *Psychology and Behavioral Sciences* 5.4 (2016) 83. https://doi.org/10.11648/j.pbs.20160504.12.

Zuckerman, M, et al. "Sensation Seeking Scale." sjdm.org, 1978. https://sjdm.org/dmidi/Sensation_Seeking_Scale.html.

Zweig, Connie. "Ken Wilber's Call to Grow Up, Clean Up, Wake Up, and Show Up." *Psychology Today*, Dec. 3, 2021. https://www.psychologytoday.com/ca/blog/shifting-role-soul/202112/ken-wilbers-call-grow-clean-wake-and-show.

www.ingramcontent.com/pod-product-compliance
Lightning Source LLC
Chambersburg PA
CBHW050332230426
43663CB00010B/1828